AGRICULTURE AND THE ENVIRONMENT

ELLIS HORWOOD SERIES IN
ENVIRONMENTAL MANAGEMENT, SCIENCE AND TECHNOLOGY
Series Editor: Dr YUSAF SAMIULLAH, Principal Environmental Specialist, ACER
Environmental, Daresbury, Cheshire
Consultant Editor: Professor R. S. SCORER, Emeritus Professor of Environmental Science,
Imperial College of Science and Technology

Bache and Johnstone	**MICROCLIMATE AND SPRAY DISPERSION**
Cluckie and Collier	**HYDROLOGICAL APPLICATIONS OF WEATHER RADAR**
Currie and Pepper	**WATER AND THE ENVIRONMENT**
Haigh and James	**WATER AND ENVIRONMENTAL MANAGEMENT: Design and Construction of Works**
Howells	**ACID RAIN AND ACID WATERS**
James and Eden	**ENGINEERING FOR PUBLIC HEALTH**
Kay	**RECREATIONAL WATER QUALITY MANAGEMENT: Vol. 1 Coastal Waters**
Kay and Hanbury	**RECREATIONAL WATER QUALITY MANAGEMENT: Vol. 2 Fresh Waters**
Kovacs, Podani, Tuba, Turcsanyi, Csintalan and Menks	**BIOLOGICAL INDICATORS IN ENVIRONMENTAL PROTECTION**
Price-Budgen	**USING METEOROLOGICAL INFORMATION AND PRODUCTS**
Priede and Swift	**WILDLIFE TELEMETRY: Remote Monitoring and Tracking of Animals**
Ravera	**TERRESTRIAL AND AQUATIC ECOSYSTEMS: Perturbation and Recovery**
Scorer	**CLOUD INVESTIGATION BY SATELLITE**
Scorer	**SATELLITE AS MICROSCOPE**
Scorer	**METEOROLOGY OF AIR POLLUTION: Implications for the Environment and its Future**
Simpson	**GRAVITY CURRENTS: In the Environment and the Laboratory**
Szabo	**RADIOECOLOGY AND ENVIRONMENTAL PROTECTION**
White, Bellinger, Saul, Symes and Hendry	**URBAN WATERSIDE REGENERATION: Problems and Prospects**

AGRICULTURE AND THE ENVIRONMENT

Editor:
JOHN GARETH JONES,
Director of Science and Quality, Wessex Water, Bristol

ELLIS HORWOOD
NEW YORK   LONDON   TORONTO   SYDNEY   TOKYO   SINGAPORE

First published in 1993 by
ELLIS HORWOOD LIMITED
Market Cross House, Cooper Street,
Chichester, West Sussex, PO19 1EB, England

A division of
Simon & Schuster International Group
A Paramount Communications Company

Printed and bound in Great Britain
by Bookcraft, Midsomer Norton

British Library Cataloguing in Publication Data

A catalogue record for this book is available from the British Library

ISBN 0–13–065863–4

Library of Congress Cataloging-in-Publication Data

Available from the publisher

Table of contents

Foreword

There is increasing conflict between agriculturists and environmentalists as a consequence of the development of intensive farming practices and the use of modern fertilizers and other chemicals. As a professional body, the Institution of Water and Environmental Management has endeavoured to present a balanced and pragmatic viewpoint on all environmental issues, free from emotion and drama.

Silage liquor is hundreds of times more polluting than crude sewage! The use of agricultural chemicals is seriously affecting flora and fauna! And yet, it would be quite unacceptable to return to the old method of farming and expect the UK to achieve the productivity needed to meet the needs of more than 50 million people. What is the present situation, and what guidelines are there for the way forward into the next century?

This Symposium attempts to provide answers to these questions by bringing together papers from authors on both sides of the argument. There are benefits from the use of modern fertilizers and agricultural chemicals. Fish farming can make a valuable contribution to food protein demands. There are ways and means of ensuring that nature is not only protected but actually enhanced by consideration of conservation in conjunction with farming.

The proceedings of the Symposium set more balanced perspectives than those which seem to be most widely seen in the daily papers. The best solutions to problems will never come unless opposite sides of the argument are recognized and discussed, so that views do not become entrenched in blind bias.

Delegates of various disciplines and of wide-ranging interests were brought together at the Symposium, and it is hoped that the proceedings provide a better understanding and appreciation of different interests, and of how they might best be served in the future.

Farmers, chemical manufacturers, conservation interests, and the water industry need to meet and decide how their interests will be best served for the future.

Geoffrey Truesdale
IWEM Past President

Chapter 1

Modern agriculture relies heavily on the sensible use of agrochemicals to ensure the economic production of crops. The quality of crops and their customer acceptability are also dependent upon wise agrochemical management. The proper use of safeguards ensures that environmental impact is kept to an absolute minimum. In some cases, treatment of agrochemical residues is required. Processes have been simplified and packaged for use on farms to treat the washings from spraying machines, which will prevent the pollution of water resources.

Animal manures including sewage sludge depend on agricultural land as the principal outlet for recycling. Practices developed over the last 30 years for the disposal of sludges have been designed to maximize the benefit to farmers, while at the same time controlling potential issues of public nuisance and water pollution. Regulation of the disposal of sewage sludge to land has further secured environmental protection.

Paper 1   **The farmer's need for agricultural chemicals**
J. Malcolm, Head of Arable Department,
National Farmers' Union

Paper 2   **Prevention of water pollution in manufacture
and use of agrochemicals**
K. S. Johnson, East Peckham Environmental
Consultants and D. A. Harris, Sterling Services

Paper 3   **Agricultural utilization of sewage sludge**
R. D. Davis, Manager, Land Disposal and
Environmental Toxicology Group, Water Research Centre

Discussion on Papers 1, 2, and 3

Authors' replies

1

The farmer's need for agrochemicals

J. Malcolm, TD, BA (Oxon), MSc (Econ)[†‡]

ABSTRACT

This paper examines the use of agrochemicals in modern agriculture and horticulture. It begins by reflecting on the uncertainties of agricultural and horticultural production before the use of synthetic chemicals, then discusses the improved quantity and quality of output which are possible when using agrochemicals. The social, environmental, and agricultural hazards attendant on chemical use are considered. The paper concludes that, always provided the proper safeguards are observed, the use of agrochemicals is a *sine qua non* for the great majority of Britain's farmers and growers.

KEY WORDS   Pest Disease Fertilizers Pesticide Organic farming Agrochemical Integrated pest management

BACKGROUND

During production, farmers and growers wage a constant battle against the elements and against pests and diseases which threaten their crops and livestock. The vagaries of climate and the ravages of disease and pests historically led to periods of famine interspersing times of plenty. In most developed nations, the risk of serious production losses from pests or diseases has now been all but eliminated through chemical control methods combined with good husbandry. However, in all too many countries pests and disease still cause savage losses in production and during post-harvest storage. It is estimated that worldwide some 30% of agricultural and horticultural production is lost through pests and disease.

It is worth recalling at the outset that the use of chemicals in pest, weed, and disease control is not particularly new. Farmers have long used remedies based on naturally occurring chemicals. Thus, Homer mentions the use of sulphur as a fumigant as long ago as 1000 BC. In 100 BC the Romans used the plant hellebore in controlling rats and mice, and by AD 900 the Chinese were using arsenic to control garden pests. However, it is only in the twentieth century that chemical control methods have blossomed.

[†] Head of Arable Department, National Farmers' Union.
[‡] I am greatly indebted to my colleague Dr Alastair Burn for his extensive comments on the initial draft of the paper. Any errors or omissions that remain are my own.

The development of DDT as an insecticide 50 years ago was particularly important in triggering the era of agrochemicals. The world agrochemical market grew rapidly in the post-war years, and as late as the 1970s it continued at an average rate of above 6 per cent a year. The 1980s have seen a substantial slowing in the growth rate as the overall agrochemical market now approaches maturity. As we gain in sophistication we may expect the proportion of the World's crops, land, and livestock treated with chemicals to continue to increase somewhat, though modern developments suggest that new chemicals will be applied at substantially lower dosage rates than those they replace. The development of synthetic chemicals during the twentieth century has greatly increased the number of pests and diseases that can be brought under control by chemical means. Synthetic chemicals are now widely used to prepare the ground, to regulate growth, to facilitate harvest, and to safeguard post-harvest storage. The stock farmer is similarly able to protect his livestock through a range of preventative and curative medicines. Nor should we forget that synthetic pesticides are also used on a large scale by gardeners, local authorities, public utilities, and amenity horticulturalists as well as part of land management programmes on many nature reserves.

The advantages of chemical controls are that they are rapid, economic, flexible, readily available, and easily applicable control mechanisms which, correctly used, can be highly effective in reducing pest populations to the very low levels demanded by the retailers and consumers. The disadvantages are that the manufacture depends on non-renewable energy sources, they may fail owing to pest resistance, they may cause environmental pollution or secondary pest outbreaks, and, if wrongly used, can be a hazard to operator, consumer, and even the casual passer-by.

AGRICULTURAL OUTPUT AND PRODUCTIVITY

We should perhaps remind ourselves that the objective of agriculture and horticulture is to produce foodstuffs and non-edibles, such as ornamentals, in the quantity and of the quality that the customer requires. In volume terms the expansion of UK agricultural output in the past fifty years has been enormous. In 1938, British agriculture employed more than one million people who together produced only a third of the food needs of a total population of 48 million. In 1988, the equivalent of some 450 000 full-time farmers and workers are engaged in the industry, producing three-quarters of the indigenous type food consumed by a nation of 57 million. During the past half century output per head in Britain's agriculture has increased at roughly double the rate of productivity improvement in the rest of the economy. Since the war the average milk yield from the nation's dairy herd has more than doubled; average sugar beet yields have nearly doubled, and those of wheat have increased by above 150 per cent.

A whole variety of factors have been responsible for this development—the mechanization of farming; improved education and training of those engaged in the industry; better husbandry; higher yielding crop varieties; and improved breeding of animals. By no means least among the factors causing the higher yields has been the development of agrochemicals including fertilizers, pesticides, and growth regulators. Indeed, conventional modern farming has come to depend upon agrochemicals. Modern cereal varieties rely on a high nitrogen input to reach their full yield potential. To stop unwanted stem length, growth regulators have to be applied. In some cases high nitrogen usage increases the susceptibility of the crop to pests and diseases, thus requiring a pesticide spray programme.

THE BENEFITS OF CHEMICALS TO FARMERS

We do not have to look to the past for examples of the value of agrochemicals through combating such diseases as potato blight—the cause of the Irish famines in the mid-nineteenth century. The reappearance of plagues of locusts in Africa after more than two decades of successful chemical control is primarily due to political factors preventing treatment of the sites where locusts breed, and it serves to remind us of the value of correctly timed chemical action.

Moreover, in assessing the value of agrochemicals to the economy as a whole, we must also take account of the effects they have had in releasing man from unpleasant tasks. For example, in sugar beet production before the advent of herbicides, large numbers of workers were engaged simply in hoeing between the plants and in singling out beet seedlings. The countryside no longer has the necessary manpower nor—despite high levels of unemployment in the UK—could farmers today attract the necessary numbers of workers to spend their time in such soul destroying tasks as weeding. Expectations have changed, and a widespread reversion to the methods of the past is simply not possible.

As late as 1960 the entire sugar beet crop was still hand-hoed; today, only a small part of the crop has any hand weeding—beet growers now rely on herbicides for their weed control. Since 1960, the proportion of the crop which is precision drilled has risen from about 20 per cent to 100 per cent, and this itself has created new chemical pest control requirements as each plant must reach maturity to ensure optimum yield. In short, the high cost of labour and the unwillingness of modern society to put up with the drudgery of the past has given a major stimulus to the increased need for agrochemicals in the UK.

Synthetic nitrogen fertilizer is now relied on primarily because its nutrients are readily available to the plant and hence there is a rapid response after application. Other factors are the ease of application which allows a closer match between the application rate and the nutritional demands of the growing crop than is the case with organic fertilizers. Moreover, the specialization of modern farming means that the major organic fertilizer—farmyard manure—is not readily available to the intensive arable crop grower of the eastern half of Britain. To reduce the availability of agrochemicals could therefore have a major impact on the activity and indeed viability of many farm businesses.

COST EFFECTIVENESS

Having emphasized the benefits of chemical farming I must not imply that the average farmer simply engages in a wholescale programme of spraying, dipping, injecting, or otherwise delivering chemicals onto or into his crops and livestock. With chemical pest and disease management, as with any input use, the essential criterion will be an economic one. Is the revenue lost from the impact of the pest or disease greater than or less than the cost of using chemicals to combat the problem?

There will be an 'economic threshold' below which action should not be taken but above which crop or livestock losses exceed control costs. Regrettably, there is evidence that some farmers are not using chemical pest control measures in the most economically practical way. Taking the example of cereal aphids, some half a million hectares (out of a total cereal area of 4 million hectares) are sprayed annually against this sporadic pest. Sprays are sometimes applied too soon, or too late, in the aphid population cycle, and so, despite chemical treatment, the crop still suffers a degree of aphid damage. Even taking into account the economic advantages of applying both a fungicide and aphicide spray together, growers in the UK who do not use the principle of economic thresholds or the pest forecasting systems which are available gain only part of the possible benefit available through aphid spraying.

Pest and disease forecasting and monitoring systems are becoming increasingly accurate and more widely available—but research into this area in the UK is already threatened by shortages of cash as a consequence of the Government's cutbacks in agricultural and horticultural research and development. 'Blitecast' for potato blight, and similar forecasting systems for cereal, oilseed rape, and sugar beet pests and diseases, all aim to allow a more rational use of pesticides—both reducing the input of pesticides and increasing the economic advantage to the grower.

However, the need to achieve high quality produce, rather than high yield *per se*, means that it is often not possible to operate anything other than a prophylactic treatment regime. In the case of carrot fly, for example, the risk of customer rejection of even a low infestation of carrots by maggots, means that it is not realistic to reduce the number of spray applications in large scale crop protection.

As a result of the last half century of progress, the quality of our foodstuffs and the variety of farm produce offered to our customers is greater than it has ever been. Consumers have become accustomed to the consistent quality of modern farm produce in terms of its size, shape, and appearance, and in an age of convenience cooking our customers will not readily accept blemished, deteriorating, or insect infested produce.

We would do well to remind ourselves that although there is a demand for organically grown produce, that demand is currently tiny in relation to the market as a whole, accounting for less than one per cent of total consumption. Even the most optimistic studies of the market put the demand for organic produce at still well below 10 per cent by the turn of the century. Organic production yields are substantially lower than those achieved by conventional (chemical) methods, hence growers need to achieve a significant price premium if organic farming is to be made to pay.

Of course chemical controls are not the only options—in many cases crop varieties are bred for resistance to particular pests and diseases; in some environments such as glasshouses, natural predators can be used as part of an integrated pest management approach. Environmental protection is axiomatic in this approach, which also encourages the role of native natural enemies in many outdoor crops, ranging from cereals to fruit trees.

PROBLEMS OF 'CHEMICAL FARMING'

Chemical pest and disease management has its own attendant risks and problems. Pesticides may have phytotoxic side effects, they may cause damage to nearby crops, animals, or natural habitats if applied wrongly or carelessly. Pesticides may harm the environment: some have shown persistence in rivers and water courses, some indeed in the food chain. Furthermore we must consider the possibilities that even non-persistent chemicals may have adverse long-term effects if overused.

Some of the potential hazards have long been recognized: thus herbicide residues have been known to affect subsequent crops; and the danger to bees, and hence to crop pollination, of spraying insecticides on flowering crops is well known. Other problems are only now becoming significant: thus aldrin and dieldrin have been extremely effective in combating soil-borne insects, but for many years their persistence in the soil and the adverse effects which they can cause when they entered the food chain were not realized.

The unforeseen impact of nitrogen leaching into rivers and water courses is now well recognized as causing a problem for the water industry. Artificial fertilizer is not the sole, nor in some cases the main, source of the trouble, much of which stems from the ploughing up of permanent pasture in the desperate efforts to increase home food production during the war and early post-war years. In this context it is worth noting that a widespread reversion to 'organic' sources of fertilizer would not of itself alleviate the problem of nitrogen leaching into water supplies—indeed, if associated with a switch to spring cropping, the position could worsen.

Instances of pest resurgence, or secondary pest outbreaks—as when an insect's natural enemies are more susceptible to a particular insecticide than the insect itself, are well known. For example spider mite is now a pest of apple trees, following routine spraying of those trees against other pests, and the consequent killing of predators which would have suppressed spider mite populations. Moreover, there is always the fear of the development of a resistant strain (as for example the peach–potato aphid now resistant to a very wide range of chemical control measures and posing a real threat to the sugar beet industry).

THE FUTURE

The farmers' need for agrochemicals is likely to continue for the foreseeable future. However, the complexities and high costs of registering new products to meet stringent safety requirements coupled with a limited patent life, may make the relatively small UK market less attractive to major pesticide companies. Perhaps we should expect rather fewer pesticides being developed specifically for the UK market in the future. Instead we may find 'pest management packages' becoming of greater significance.

Monitoring programmes may become more widely available in future, as may the wider application of newer control methods such as insect viruses and bacterially derived toxins—but again, we must note that there is a threat to research funds.

Despite the valuable part that the use of agrochemicals has played in the past success of British agriculture and horticulture in raising the quantity and quality of domestic production, there are increasing pressures on the industry to reduce its dependence on agrochemicals. In part this pressure is a direct result of the achievement of British farmers and growers—and their counterparts elsewhere in the developed world—in making food shortages a thing of the past. In times of plenty the public and politicians all too quickly focus their attention on the costs of storage and disposal of surplus crops. A reduced use of inputs—notably fertilizers—is held to be a comparatively painless way of eliminating the unwanted surpluses, hence the call for nitrogen quotas or other forms of restriction. However, this approach often ignores the increase in unit costs which would inevitably follow from any sub-optimal use of inputs.

The rising awareness of residue problems, be it nitrate levels in water or pesticide residues in food and water, is adding to the calls for restrictions in the volume of agrochemicals used. These pressures will undoubtedly increase, and we can expect that water quality standards are more likely to be tightened than relaxed as the monitoring and detection methods improve. Indeed we are already seeing a range of Government actions such as the proposed establishment of water protection zones, the implementation of the European water quality standards, and the introduction of more rigorous standards for the approval and use of pesticides under the UK's Food and Environment Protection Act.

Of course we must ensure that the highest possible safety standards are set so as to protect the operator, the consumer, and the environment, but we must keep the issues in perspective. Over the past fifty years the benefits to society from the use of agrochemicals have vastly outweighed the costs. Properly handled, agrochemicals should continue to play a vital role in food production.

CONCLUSION

To maintain a high level of agricultural productivity and to continue to provide the high quality, readily available produce which the retailer and consumer demands, British farmers will continue to rely on agrochemicals. However, we must recognize that there is a continuing groundswell of opinion which voices legitimate concern about the long term effects of our dependence on agrochemicals. The environmental and natural food lobbies are likely to strengthen, and with them restrictions on the farmer's freedom to use chemicals. The Food and Environment Protection Act and the regulations it has introduced into the use of pesticides will help to improve the safe use of pesticides. Nonetheless, as with democracy, the price we must pay for the benefits of agrochemicals is eternal vigilance!

BIBLIOGRAPHY

[1] Green, M. B. (1976) *Pesticides: boon or bane?* Elek Ltd, London.
[2] Finney, J. R. (1988) World crop protection prospects: demisting the crystal ball. *BCPC Brighton Crop Protection Conf.*, **I**, 3–14.
[3] HMSO (1968) Earlier data. In: *A century of agricultural statistics—GB 1866–1966.* HMSO, London. For latest data see HMSO (1989) *Agriculture in the UK.* (1988) HMSO, London.
[4] Jepson, P. C. & Green, R. E. (1983) Prospects for improving control strategies for sugar beet pests in England. In *Advances in Applied Biology*, **VII**, 175–249.
[5] Watt, A. D., Vickerman, G. P. & Wratten, S. D. (1984) The effect of the grain aphid *Sitobion avenae* (F) on winter wheat in England: an analysis of the economics of control practice and forecasting systems. *Crop Protection*, **3**, 209–222.
[6] Finch, B. (1987) Horticultural crops. In *Integrated pest management.* (eds Burn, Coaker & Jepson) Academic Press, London, 257–293.

[7] Vine, A. & Bateman, D. I. (1981) *Organic farming systems in England and Wales—practice performance and implications.* Department of Agricultural Economics, University of Wales, Aberystwyth.

[8] Burn, A. J., Coaker, T. H. & Jepson, P. C. (eds) (1987) *Integrated pest management.* Academic Press, London.

[9] Edwards, C. A. (1970) *Persistent pesticides in the environment.* CRC Monoscience Series.

[10] Jepson, P. C. (1988) Ecological characteristics and susceptibility of non-target invertebrates to long-term pesticide side effects. *Field Methods for the Study of Environmental Effects of Pesticides.* BCP Monographs, **40**, 191–198.

[11] Solomon, M. G. (1987) Fruit and hops. In *Integrated pest management.* (eds Burn, Coaker & Jepson), Academic Press, London, 329–360.

[12] Dewar, A. *et al.* (1988) The rise and rise of the resistant aphid. *British Sugar Beet Review,* **56**, (i), 40–43.

2

Prevention of water pollution in manufacture and use of agrochemicals

K. S. Johnson, MIWSoc, MInstWM, AIOH[†] and D. A. Harris, CEng, MIMechE[‡]

ABSTRACT

The treatment of effluents from the formulation and manufacture of agrochemicals by using a process of chemical flocculation and carbon filtration has been effective for many years. This process has now been simplified and packaged for use on farms to treat the washings from spraying machines and unused dilute spray liquids. A simple and inexpensive plant has been designed to use the process, and this is being evaluated in a number of countries where concern about environmental protection is high.

KEY WORDS Water pollution Effluent treatment Agrochemicals Flocculation Carbon adsorption 'Carbo-Flo' 'Sentinel'

INTRODUCTION

The treatment of effluents from the manufacturing and formulation of pesticides has been a matter of routine for nearly two decades. Pesticides in general have a low solubility in water and are usually manufactured in the form of wettable powders, suspension concentrates, or miscible liquids. The chemical treatment of effluents containing suspensions and emulsions causes flocculation and settlement of a high proportion of the pesticide ingredient in the form of a dense sludge. Further treatment of the clarified supernatant through activated carbon removes the pesticides in solution. The process offers a convenient method of detoxifying aqueous plant effluents, and it has recently been simplified and packaged to allow it to be used on farms by non-specialist farm staff.

EFFLUENTS FROM MANUFACTURING PROCESS

Effluents arise from all pesticide manufacturing processes owing to the need to frequently clean mixing vessels or other items of process plant. Spillages or surplus liquors have

---

[†] Consultant, East Peckham Environmental Consultants (EPEC). [‡] Consultant, Sterling Services.

also to be disposed of safely and efficiently. A large formulation plant may need to treat 100 m³ of effluent per day which could contain up to ten different active ingredients at concentrations of up to 1% in addition to other agents used in formulation.

THE TREATMENT PROCESS

It is necessary to collect plant effluents in a common sump. This offers storage capacity and a means of buffering the changes in composition of effluent which frequently occur.

The flocculation process is normally conducted in a batch treatment system, using a conical-based cylindrical vessel. This allows the settlement of sludge residues in the cone and a means of decanting the upper supernatant clarified effluent.

Optimum conditions for flocculation/clarification of effluent are pH 10–12. The addition of iron salts and lime (calcium hydroxide) to aqueous pesticide effluent normally induces rapid flocculation of all suspended solids. The inclusion of a small amount of polyelectrolyte will serve to accelerate the coagulation of flocs and subsequent settlement. The addition of an adsorbent clay and possibly powdered activated carbon can be beneficial in the removal of trace residual pesticide if this is deemed to be necessary.

Dosage rates of chemical flocculants need to be established for specific effluents, but the following can serve as a starting point.

	mg/litre
Lime pH 11–12	500
Ferric sulphate (40% solution)	200
Polyelectrolyte (anionic type)	5
† Powdered clay	1000
† Activated carbon	500

† Carbon and clay are normally dosed as a secondary stage following initial flocculation and separation/removal of sludge.

A vessel of 5–10 m³ capacity with mechanical agitation is ideal for collection and treatment.

To maintain a consistent standard of high quality finish effluent, a second stage of activated carbon adsorption should be considered. This should comprise at least 2×1 tonne beds of high activated carbon granules (14/44 mesh), surface area > 1000 m²/g.

Coal or wood-based carbon granules will give the best adsorption performance. Coconut shell carbons, by virtue of their relatively small pore size, are generally unsuitable for pesticide effluent treatment.

Flow rates of clarified sand-filtered effluent should be regulated through the beds to allow a minimum of 1 hour contact residence time.

Final effluents are normally clear, virtually colourless, and non-toxic. Disposal can be directed to a sewer, soakaway, or in hot climates possibly an evaporation pond. Direct discharge of effluent to a water course is *not* recommended.

Sludges from the process can be dried in shallow drying beds and subsequently disposed of to a designated waste disposal site. The sludges are normally maintained at pH 12–14 during storage to accelerate the degradation of pesticide residues.

ICI Agrochemicals developed this process in 1970 and has used it at its main formulation plant in the UK since 1972. It is now in use in 20 medium sized formulation plants worldwide.

THE APPLICATION OF AGROCHEMICALS

Agrochemicals are usually diluted with water for application to soil or to growing crops as a spray. In recent years the spraying equipment available to farmers has become highly sophisticated with effective instrumentation and controls to allow close control of the application rate. Enclosed cabs on tractors, hydraulic operation of booms, low level filling devices, and improved mechanical design have all contributed to keeping the contamination of the operator to a minimum. Automatic control of application rate in proportion to forward speed, better spray nozzles and boom suspensions, and growing techniques which use 'tramlines' to control the passage of all field machines in a crop, have all contributed to more precise application rates and thus to the reduction of pollution during normal field use. The spray operator is still a vital link in the chain, however, and only he can ensure that spraying does not take place in unsuitable weather and that water courses, for example, are not accidentally sprayed. Again, recent developments in the legal framework for spraying and the training and certification of operators are making a contribution to improving standards.

The one area in which little progress has been made is in the disposal of unused spray material and the dilute contaminated water which is produced when sprayers of all sizes and types, from knapsacks to aircraft, are washed out. Even if precise quantities have been calculated and mixed, unused material is always present in the tank, pump, and boom at the end of a spraying job, and, depending on the chemical to be applied next, this will have to be disposed of and the machine thoroughly cleaned. In intensive growing situations, cleaning may be needed several times per day.

Growers used to be advised to spray out dilute washings onto stuble or other fallow land, but this was rarely available, and the time required was not easily found. There is a hiatus in this area because of the unacceptability of the advice contained in the draft code of practice for agricultural spraying which proposed application to a sacrificial area, securely fenced and away from all water courses and catchment areas.

In the USA it is common for washings and unwanted spray strength liquid to be stored in large tanks and used as diluent for subsequent applications of the same type of agrochemical. This is expensive in tankerage but is much less suitable for UK conditions because of the more complex cropping systems and the multiplicity of chemicals in use.

THE TREATMENT OF EFFLUENT ON THE FARM

A new approach, launched at the end of 1987 after several years of development, is a simplified version of the chemical treatment process used in manufacturing plants described above. The name 'Carbo-Flo' has been applied to the process (Fig. 2.1), and a dedicated small scale plant called 'Sentinel' was developed to use it.[†]

[†] 'Carbo-Flo' is a trademark of Imperial Chemical Industries PLC, 'Sentinel' is a trademark of E. Allman & Co. Ltd.

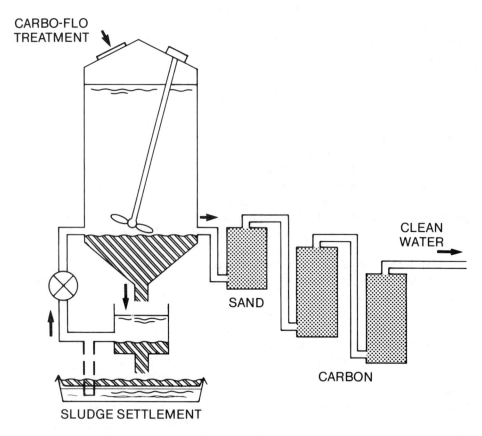

Fig. 2.1. ICI 'Carbo-Flo' process.

The process employed is identical to that of the large scale factory plant, but the treatment chemicals have been pre-packed to match the 1 m³ batch size of the 'Sentinel'. Packs contain sufficient material to treat washings and dilute sprays up to about half the recommended application rates (*c.* 1000 ppm a.i.). They are supplied in four plastic bottles clearly labelled for sequential addition to the effluent. A marker dye is incorporated to act as a visual indicator of the life of the carbon filters.

The 'Sentinel' plant is compact (1 m × 2 m × 3 m) and is designed for outdoor use. It is portable by various means, or can be fixed. It has a 1000 l (1 m³) flocculation tank, a sludge concentration system, and a sand filter to protect the two activated carbon columns which are connected in series. The capacity of each carbon filter is approximately 20 m³, so, when the marker dye shows through the first filter and it is time to fit a new one, there is still a large adsorption capacity available. It is expected that a typical farm would need to change a filter only once per year, although this will depend on the intensity of use.

The 'Sentinel' unit requires attention for only the first phase of treatment when the chemicals are added and the agitator is in operation, and then some 30 minutes later when flocculation has taken place and the outlet valve is opened to allow clarified liquid to flow through the carbon filters. Sludge need be treated only every 3 or 4 batches. This is a two-stage process resulting in a dense wet sludge which can be dried down to a solid for disposal. The current plant has been designed for manual operation to keep costs as low as possible. Future models are expected to have automatic time controls, motorized valves, and chemical loading devices available as options. A special model for the treatment of sheep dip is under consideration, and the need for plants in a wide range of sizes is being assessed.

Table 2.1. 'Carbo-Flo' effluent treatment process effluent quality; summary of analytical results (mg/l)

Treatment stages	TOC	'Lindane' HCH	'Captafol'	'Paraquat' ION	Pirimicarb
1. Untreated	603	11	75	180	700
2. Chemical clarification	490	3	3	12	430
3. Carbon treated	30	< 0.001	< 0.01	< 0.01	< 0.02

The following pesticides, detected in farm spray-tank effluents, were all removed by stage 3 activated carbon to <0.01 mg/litre: Trifluran, Propiconazole, Iprodine, Di-allate, Fluazifop-butyl, Fluroxy Pyr-(1-Methyl Heptyl), Methyl-MCPA.

The disposal of treated effluent and the sludge and carbon filter elements will vary from country to country and according to the requirements of the responsible water authorities. Discussions to establish consent procedures for the disposal to soakaways or foul sewers are taking place in the UK. Further research into quality of the treated effluent (Table 2.1) and the composition of typical sludges is taking place in a number of countries as a preliminary to local marketing. In general the reception of this simple concept has been enthusiastic in countries where there is a heightened awareness of the risks of point source pollution from intensive agricultural operations. It is seen as a worthwhile contribution to the prevention of pollution in an area of activity in which no other good solution is available.

3

Agricultural utilization of sewage sludge

R. D. Davis, BSc, PhD (Member)[†]

ABSTRACT

Utilization on agricultural land is the principal outlet for sewage sludge in the UK, accounting for some 40% of annual production. Current practice has developed over the last 30 years to maximize the benefits to farmers while controlling potential problems of public nuisance, water pollution, pathogen transmission, and soil contamination. Until now this has been achieved by Government guidelines, but in future, use of sewage sludge in agriculture will have a statutory basis after implementation of an EC Directive in June 1989.

KEY WORDS   Sewage Sludge   Agriculture   Regulations   Fertilizer value   Contaminants   Pathogens.

INTRODUCTION

Water re-use and the protection of river quality depend on the collection and treatment of sewage. In the nineteenth century inland treatment often involved broad irrigation of crude sewage over dedicated land, but this soon became inadequate to deal with the volumes of sewage being produced. The sewage farms were replaced by sewage-treatment works (STW) designed to treat municipal sewage by separation of settleable solids (primary treatment) and conversion of dissolved and colloidal solids by a biological process (secondary treatment) to metabolites such as carbon dioxide, sulphate, and water and also microbial cells and residues in a flocculent form (humus or surplus activated sludge) which can be separated by a secondary stage of settlement. The end products are a comparatively clean effluent which can be returned to the river either directly or after some further (tertiary) treatment, and a sludge containing much of the organic load of the original raw sewage. Sludges from the primary and secondary treatment processes are usually combined before disposal, the rates of primary sludge solids to secondary sludge solids often being about 2:1 by weight. Sewage sludge is a putrescible, thin slurry typically with a solids content of about 2% of which 70–80% is organic matter. From the point of view of disposal, the advantage of treating sewage in

[†] Manager, Land Disposal and Environmental Toxicology Group, Water Research Centre.

this way is that the volume of sludge produced will only about 1% of the volume of raw sewage received, and this will be further reduced by the thickening and treatment of the sludge which is usual before disposal. Nevertheless, sludge treatment and disposal remains a problem for the water industry and costs about £250 million per annum or about half the total cost of sewage treatment.

Until the late 1950s it was common to consolidate sludge on drying beds which it was spread in heavy dressings on adjacent land dating back to the sewage farm days or stockpiled at the STW and given to farmers or others prepared to take it. Sludge utilization on dedicated sites owned by the water utilities continues to be practised successfully in some locations, but UK practice in the 1980s is largely based on the supply and transport of sludge by the utilities to farms within a 20 km radius of STWs where it is spread in accordance with crop requirements for nutrients at rates of 5–10 tonnes dry solids (tds)/ha.

The practice of taking the sludge from the STW to farmers was first developed in the late 1950s at some of the larger STW near to conurbations where sludge disposal was becoming an increasing problem. A typical example is the scheme built up by the West Hertfordshire Main Drainage Authority for the use of liquid digested sludge from Maple Lodge STW (now part of Thames Water). Kershaw *et al.* [1] have described how the system was started in 1956 with the purchase of two old RAF fuel tankers of 4.5 m^3 capacity. The West Hertfordshire operation developed so successfully that the volume of sludge going to land increased from 3200 m^3 in 1957/58 to 111 000 m^3 in 1960/61. By then 19 tankers were in use and they covered a distance of 400 000 km averaging 12 km per load, the total volume of sludge being distributed over 1000 ha at an average of 110 m^3 wet sludge/ha. Utilization of sludge in this way was considered to be by far the cheapest and most efficient disposal method for Maple Lodge STW. Throughout the 1960s other operations followed suit, and the supplying of sludge to farms became the main method of sludge disposal for inland STW. In 1970 a Government report on sludge disposal [2] included the recommendation that wherever possible encouragement should be given to the application to agricultural utilization on land was a major disposal outlet, and it was evident that guidelines were needed to ensure that operations were carried out satisfactorily.

GUIDELINES FOR SLUDGE UTILIZATION

Development

In 1971 MAFF published an advisory note [3] intended to deal with contamination problems in sludge-treated soil. It set out recommendations for zinc, copper, nickel, and boron. The Department of the Environment put the guidance on a broader basis with the establishment in 1974 of the Working Party on the Disposal of Sewage Sludge to Land, and after reorganization of the water industry at that time, the Standing Committee on the Disposal of Sewage Sludge. This committee was to be a national focus of expertize to advise generally, to recommend, assess, and if necessary to monitor research and development and to publish guidance on good practice. A sub-committee set up to deal with the disposal of sludge to land agreed guidelines which were published in 1981 [4]. As all interests had been represented on the sub-committee, this comprehensive set of guidelines found widespread acceptance and is the basis of current UK practice. The Department of the Environment will update existing in 1989 in a Code of Practice on the Use of Sewage Sludge in Agriculture [5] which is designed to ensure that when sludge is used in agriculture:

(1) There is no conflict with good agricultural practice.
(2) The long-term variability of agricultural activities is maintained.
(3) Public nuisance and water pollution are avoided.
(4) Human, animal, and plant health are not put at risk.

Apart from taking account of new developments in research, the update is needed to include the statutory requirements of the Council of the European Communities Directive on the protection of the environment, and in particular of the soil when sewage sludge is used in agriculture [6, 7]. The Directive presumably arose from the Commission's wish to harmonize guidelines for sludge utilization throughout Europe, since most Member States already have domestic regulations, and reflects the facts the fact that agricultural land is the outlet for 34% (1.88 million tds per annum) of EC sewage sludge [8]. The Directive takes a positive view in agriculture in such a way as to prevent harmful effects on soil, vegetation, animals, and man, while encouraging its correct use.

Value of sludge to the farmer

Use of land is of greater importance to the water industry, which uses the outlet for 40% of its annual sludge production, than it is to agriculture, since only 1–2% of farmland benefits from sludge spreading each year. Also, sludge utilization accounts for only about 1.2% of the nitrogen input to agricultural land (for P, 2.2%) from chemical fertilizers and housed livestock manure and thus makes little impact on the national fertilizer budget. However, the combined value of the available nitrogen and phosphorus content of sludge is about £5.8 million per annum, so sludge is potentially valuable for the farmer who uses it. To encourage him to do so, the water utilities seek to provide a professional service detailing the type of sludge to be supplied and defining its fertilizer value as far as is possible for an organic material.

The new Code of Practice requires sludge to be applied at rates which take account of crop requirements for nutrients. Sludge analysis is carried out to determine the content of phosphorous and total and ammoniacal nitrogen, these details being supplied to farmers on request. Ammoniacal nitrogen in liquid digested sludge is readily available for crop uptake, and gives a rapid response, particularly when applied to grassland. Nitrogen in undigested and unwatered sludges is released more slowly, but the extent of release can be defined by crop trials comparing response to sludge with response to the inorganic fertilizers familiar to farmers. Over the years, extensive field trials have been carried out to define the fertilizer value of different types of sewage sludge [9–11], taking account of soil texture and geographical location. Most of the work has used grass or cereals as the indicator crop, but studies include work on, for example, potatoes and peat soils [12]. The emphasis in all this work has been on nitrogen, but sludge is a rich source of phosphate in terms of crop requirements, and the availability of its phosphate content for crop growth is less dependent on sludge type [13]. Sludge contains small amounts of potash in a form readily available to crops [11]. It contains calcium, especially if treated with lime, and trace elements also. Sludge organic matter improves soil physical conditions for crop growth, including aggregate stability and water holding capacity, and is as effective in this respect as the organic matter of farmyard manure [9, 14]. Practical information distilled from the results of field trials is summarized in the WRc *Farmer's guide to the agricultural value of sewage sludge* [15].

Sludge quality and method of application are also important components of the service. Farmers are likely to be doubtful about surface applications of mechanically dewatered sludge cake to grassland because of the possibility of sward damage and direct ingestion of sludge by grazing animals. On the other hand, pasture grass responds well to top dressings of liquid digested sludge, but adequate screening is necessary to remove refuse material which would otherwise remain on top of the grass or cause blockages in spreading equipment. Arable farmers will want to avoid soil compaction and excessive applications of greasy material. The new edition of the WRc *Directory of equipment* [16] includes sections on sludge storage, pumps, pipes, and fittings as well as revised sections on road tankers, field tankers, tractor drawn tankers, muck spreaders, and soil injection and irrigation equipment. There are general sections on spreading devices and tyres, with the sections on choice of tyres and traction requirements being extensively revised to provide advice on the avoidance of soil damage from heavy sludge spreading equipment (Fig. 3.1). A major development in recent years has been the emergence of subsurface injection as a method of application of sludge. Where ground conditions are suitable, with a skilled operator, this can be a very satisfactory way of applying sludge to both grassland and arable land. Since the sludge is buried beneath the soil surface there is little risk of pathogen transmission or odour problems. There is a WRc *Code of Practice on Injection* [17], and training courses for operators are held in collaboration with Silsoe College.

Fig. 3.1. Tanker spreading treated sewage sludge on grassland.

Control of public nuisance and water pollution
The EC Directive [6] states that sludge shall be treated before being used in agriculture unless it is injected or immediately worked into the soil. As described below, the definition of treated sludge makes reference to reduction in fermentability, so the aim of this restriction is to reduce the likelihood of odour nuisance as well as to control problems of pathogen transmission. Operators are well aware that odour nuisance is one of the more common causes of day-to-day complaints, and that the public nose is becoming more sensitive and less tolerant to the smells of country life. This has to be taken seriously, and apart from the statutory treatment and incorporation requirements of the Directives the *Code of Practice* describes the commonsense procedures which need to be followed in a section of environmental protection. This section draws attention also to the need for route planning in relation to both smell and traffic problems. The use of rain guns for sludge spreading is a sensitive issue, just as it is for farm slurries. In future they can be used only for treated sludge, but even so, discretion will be necessary.

The *Code of Practice* sets out guidance on avoiding surface and groundwater pollution which is a statutory requirement under Article 8 of the EC Directive. Control is based on taking account of topography, rates of application, soil conditions, and proximity to water courses and water supply sources, particularly in designated sensitive areas. Since sludge is produced continuously throughout the year it is often essential to spread it on the land in the autumn, but it appears that sludge nitrogen, being in the ammoniacal or organically bound form, is not leached as readily as fertilizer nitrogen [18]. While sludge is used on only 1–2% of agricultural land and accounts for 1.2% of nitrogen inputs, attention to conventional farm fertilizer practices and to the use of animal slurries will be a more effective way of protecting groundwater from pollution by nitrates.

Pathogens
Certain pathogens occur inevitably in sewage and in sludge, their numbers reflecting the health of the local community and the occurrence of abattoirs, for instance, in the catchment of the STW. The risk posed by pathogens in sludge has been discussed in detail by Pike & Carrington [19]. In the UK the problem of potential transmission of disease after the spreading of sludge on land is considered to relate principally to bacteria of the *Salmonella* species and eggs of the beef tapeworm, *Taenia saginata*. *Salmonella* infections are widespread in man (a major cause of food poisoning) and in food animals. Although sludge is one route by which these bacteria may be brought into the agricultural environment, it is recognized that the main routes of transmission in food animals are from animal to animal directly, or indirectly through foodstuffs and slurry [20]. Similarly, in man, the recognized cause of *Salmonella* food poisoning is faulty food hygiene in the kitchen, usually concerning meat and dairy products. Other routes than sludge exist also in the case of infection of cattle by the beef tapeworm, such as direct defaecation on land and transfer from sewage works by birds. The approach developed to successfully avoid disease transmission where sludge is used on land depends on a dual barrier system, using knowledge both of the effectiveness of sludge treatment processes to reduce the numbers or infectivity of pathogens, and of the decay rates of pathogens after application of sludge to land, which will indicate appropriate restrictions on land use.

As mentioned previously, the emphasis of the EC Directive is very much on the use of treated sludge defined as having undergone biological, chemical, or heat treatment, long-term storage, or any other appropriate process so as significantly to reduce it fermentability and the health hazards resulting from its use. Using evidence from research into the survival and infectivity of pathogens during sludge treatment [21. 22], the *Code of Practice* goes further and lists acceptable sludge treatment processes with examples of satisfactory operating conditions. Further protection against disease transmission is provided by recommended land use practices, for example for pasture grass and forage crops, and fruit and vegetable crops normally eaten raw.

Soil contamination

During sewage treatment some 70–90% of the metal load of incoming raw sewage is transferred by adsorption and precipitation into sludge. Some of these metals, for example chromium, originate predominantly from industrial discharges, while significant quantities of others, such as zinc and copper, come from products used in the home and domestic plumbing systems. Limits on metals in discharges to the sewer from industrial premises have for many years taken account of the need to ensure that sewage sludge is suitable for agricultural land [23]. Stricter controls on industrial discharges reduced the concentration of cadmium in sludge at one major STW from 145 mg/kg ds in 1971 to 18 mg/kg in 1977 [24]. This example is indicative of a general trend toward declining metal concentrations in sludge. Now the concentration of cadmium in sludge going to land is often less than 10 mg/kg ds. Taking the EC countries together, Hutton [25] has estimated the total cadmium input to arable soils in areas away from localized contamination to be 8 g/ha yr. Phosphatic fertilizer was thought to contribute 5 g/ha yr, and atmospheric deposition 3 g/ha yr. The contribution from sludge applications was considered to be too small on a national or regional basis to warrant inclusion.

Nevertheless, sludge usually contains higher concentrations of metals than most soils do, and for fields receiving it, sludge represents the main input of metals. When sludge is applied to soil, the metals tend to accumulate in the cultivated layer, and additions of sludge must be regulated so that soil concentrations of potentially toxic elements never reach levels which could be toxic either to crops or to the animals which eat them. This is a well recognized problem (limits for metals in sludge treated soil were first set 20 years ago [3], and extensive research has been carried out over the years into the effects of contaminants in sludge on soils and crops in order to provide a sound scientific basis for setting appropriate limits [26, 27]).

To ensure adequate safeguards, the new EC Directive [6] places a statutory requirement on the water utilities to monitor sludge for zinc, copper, nickel, cadmium, lead, mercury, and chromium. Rates of application of these elements to the soil have to be controlled so that soil concentration limits for the elements are not exceeded. This must be checked by sampling and analysis of the soil. The UK *Code of Practice* [5] extends these requirements to include operational monitoring of levels of molybdenum, selenium, arsenic, and fluorine. The soil limits are adjusted to take account of land use and sampling depth, and also soil pH value. The sludge producer has to keep up-to-date records including where the sludge was applied, how it was treated, the quantity applied, and details of soil and sludge analysis. The *Code of Practice* [5] makes reference to the need for vigilance concerning synthetic organic contaminants, and entry of such compounds to the sewer is restricted by the water utilities.

THE FUTURE

Agricultural use of sewage sludge is an established operation which has been carried out satisfactorily over many years on the basis of best practice according to current knowledge. Implementation of the EC Directive [6] and the new *Code of Practice* [5] in 1989 will mark a further step forward toward securing the outlet for the future. The comprehensive requirements of the legislation and the commitment of the water utilities to comply with them should help to dispel any lingering doubts about the acceptability of the practice occasionally raised by the sensational headlines which sludge can sometimes attract because of its origin, smell, and content of pathogens and contaminants. The outlet depends on the cooperation of farmers in a recycling exercise of benefit to the community at large. Improvements in sludge quality can be expected to continue through the introduction of cleaner technology and stricter limits on discharges at industrial premises, together with improved or new sludge treatment processes, for example composting. At the same time, research is continuing to improve methods of application and the scientific basis for limits and land use practices. Whether agricultural utilization will expand depends also on its cost and accessibility relative to other outlets such as codisposal in sanitary landfills and incineration. Since it is the only outlet associated with obvious environmental benefits from the recycling of plant nutrients and organic matter, it is to be hoped that utilization will remain a viable option suitable for land reclamation and forestry as well as agriculture.

CONCLUSIONS

(1) Agricultural utilization of sewage sludge is a well-established disposal option in the UK which has developed over the last thirty years.
(2) There have been voluntary guidelines to regulate agricultural utilization for about twenty years, but statutory requirements will be introduced in June 1989 by the implementation of an EC Directive.
(3) The water industry uses agriculture to receive 40% of its annual sludge production, but requires only 1–2% of farmland in England and Wales for this purpose.
(4) Farmers take sludge voluntarily, therefore sludge producers must offer them a professional service maximizing the benefits of sludge for the farmer and avoiding environmental problems.
(5) Research into the effects of contaminants, pathogen destruction and operational aspects has a continuing role in supporting regulations and the professional service and allaying unnecessary public anxiety about sludge utilization.
(6) Apart from agricultural utilization there is also scope to recycle more sludge to forest land and to land for reclamation.

REFERENCES

[1] Kershaw, M., Wood. & Ainsworth, G. (1962) *J. and Proc. Inst. Sewage Purification*, **61**, 521.
[2] Ministry of Houssing and Local Government/Welsh Office (1970) *Taken for granted.* HMSO, London.

[3]   Ministry of Agriculture, Fisheries and Food (1971) *Permissible levels of toxic metals in sewage used on agricultural land.* Advisory Paper 10, MAFF, Pinner.

[4]   Department of the Environment/National Water Council (1981) *Report of the Subcommittee on the Disposal of Sewage Sludge to Land.* STC 20, DoE, London.

[5]   Department of the Environment (1988) Use of sewage sludge in agriculture – *Draft National Code of Practice.* DoE, London.

[6]   Commission of the European Communities (1986) *Official J. of the European Communities*, **29**, L181/6.

[7]   Davis, R. D. (In press) EC Directive on Sewage Sludge, 1988. In *The changing impact of EC legislation on the water industry.* IWEM, London.

[8]   Bowden, A. V. (1987) *Survey of European sludge treatment and disposal practices.* PRU 1656–M, WRc, Medmentham.

[9]   Coker, E. G. (1983) *Wat. Sci. Tech.*, **15**, 195.

[10] Coker, E. G., Hall, J. E., Carlton-Smith, C. H. & Davis, R. D. (1987) *J. Agric. Sci.*, Camb., **109**, 479.

[11] Coker, E. G., Hall, J. E., Carlton-Smith, C. H. & Davis, R. D. (1987) *J. Agric. Sci.*, Camb., **109**, 467.

[12] Dawson, S. E. (1987) *J. Agric. Sci.*, Camb., **108**, 523.

[13] Coker, E. G. & Carlton-Smith, C. H. (1986) *Waste Management and Research*, **4**, 303.

[14] Guidi, G. & Hall, J. E. (1983) Some effects of sewage sludge physical conditions and plant growth. In *The influence of sewage sludge applications on physical and biological properties of soil.* Catroux *et al.* (eds), D. Reidel, Dordrecht.

[15] The agricultural value of sewage sludge – *A Farmer's Guide.* WRc, Medmenham.

[16] Hall, J. E. (1988) Application of sewage sludge to agricultural land – *A directory of equipment.* PRU 1857–M, WRc, Medmenham.

[17] Soil injection of sewage sludge – *A Code of Practice.* WRc, Medmenham.

[18] Hall, J. E. & Williams, J. H. (1984) The use of sewage sludge on arable and grassland. In *The utilisation of sewage sludge on land.* S. Berglund *et al.* (eds), D. Reidel, Dordrecht.

[19] Pike, E. B. & Carrington, E. G. (1986) Stabilisation of Sludge by Conventional and Novel Processes–A Healthy Future. In *Agricultural Use of Sludge–is there a Future?* IWPC, Maidstone.

[20] House of Lords (1983) Select Committee on the European Communities, *Report on sewage sludge in agriculture*, pp. 38–59, HMSO, London.

[21] Pike, E. B., Bruce, A. M., Carrington, E. G., Oliver, B. & Harman, S. A. (1988) *Destruction of pathogens in sludge.* PRD 1832–M, WRc, Medmenham.

[22] Pike, E. B. (1988) *Infectivity of taenia eggs after sludge treatment.* PRD 1199–M/2, WRc, Medmenham.

[23] Brindley, P., Carter, D. C. & Linsmith, M. J. (1982) *Effluent and Water Treatment J.*, **22**, 303.

[24] Wood, L. B., King, R. P. & Norris, P. E. (1979) Some investigations into sludge-amended soils and the implications for trade effluent control. *WRc Conf. on Utilisation of Sludge on Land.* WRc, Medmenham.

[25] Hutton, M. (1982) Cadmium in the European Community. *Monitoring and*

assessment research centre Tech Rep 26, MARC, University of London.

[26] Carlton-Smith, C. H. (1987) *Effects of metals in sludge-treated soils on crops.* TR251, WRc, Medmenham.

[27] Davis, R. D. & Carlton-Smith, C. H. (1984) *Environ. Pollut. ser. B*, **8**, 163.

Discussion on Papers 1, 2, and 3

Mr K. Guiver (Southern Water Authority), opening the discussion, said that the requirement for the use of agrochemicals must be recognized. He recalled that in the past he had
been informed that the UK would never be self-sufficient agriculturally in terms of its
productivity, but these days that was just not so. It was the use of chemicals that had
achieved this. There had been, and still were, areas in the World where the use of chemicals
brought about tremendous alleviation of disease. Examples in Europe were areas of
Southern France and Italy, where the use of chemicals had realized great benefits. Both
these countries were now partners of the UK within the EC, and it had gone from one
extreme to the other, from having a deficit to having food 'mountains'. There was a need
to sort out the agricultural policy and start taking land out of productivity. This would be
a tremendous change and would take much getting used to, particularly for the farmers
themselves.

He said that it had to be accepted that mistakes had been made. Chemicals had been
produced which had not been fully checked out before being marketed. It was only when
such occurrences arose that more thought was given to how control should be implemented.
He suggested that over the last ten years the controls that had been introduced were
satisfactory for preventing problems that had previously arisen with the use of new
chemicals. He asked for comment from Mr Malcolm on this. He asked Mr Johnson and
Mr Harris to comment on the cost of bringing new chemicals onto the market.

Mr Guiver said that there had to be safety factors. Some of the factors applicable to
Southern Europe may not seem correct in the UK, for example nitrate. He saw no scope
for changing the 50 mg/l nitrate limit regardless of concern as to whether there was medical/
toxicological back-up for it. He considered that there was no evidence of a link between
stomach cancer and nitrate concentrations—the reason for the limit was the 'blue-baby
syndrome'. However, there was a need for toxicological data to back up standards such as
those on pesticides, and he asked for the authors' thought on safety factors on drinking
water limits.

With regard to pesticide treatment, he felt that the authors had indicated a modern
treatment technique that kept pace with the development of new chemicals. He said that
the farmers' main interest was in producing food, and they would only use a foolproof
method to control the use of their chemicals. For example, anaerobic digestion on farms
had been recommended for many years, but had never really 'caught on' because of the
problems of looking after a digester. He said that the Carbo-Flo process seemed suitable
and easily operated, but asked the authors to amplify the results and to say whether there
would be further development. He felt that £4500 was still fairly expensive for farmers.

Mr Guiver asked for comment on the operation of exercises to withdraw chemicals from the market. There had been examples of problems with collection centres for farmers to hand in such chemicals.

With reference to Dr Davis's paper, Mr Guiver said that many years ago he had taken part in crop trials using sewage sludge; work that had been carried out in conjunction with the Soil Association, and excellent results had been achieved. He felt that it was a pity that such work and much other work over the years had not led to a greater use of sewage sludge. He was surprised that the conservationists did not insist on the sludge going back to land, because other methods of disposal were a waste of the nitrogen and phosphate that were in the sludge.

He said that Dr Davis had referred to a figure of £6 million and asked for clarification as to whether that referred to just the sludge that was at the moment being used on land, or whether it applied to the total sludge production in the UK?

Mr Guiver said that the two problems were (a) smell, and (b) the concern over metals and biological factors. The first could be overcome by digestion and the second by ploughing in. With digestion making the nitrogen more readily available and reducing odour, he said that he was surprised that it had not been more widely used. He was sure that the new code of practice should reduce metal problems. Industrial effluent control was already fairly comprehensive and, with cut-back in industry during the last ten years, he felt that the problem was being overcome. He did not think that there had been many instances where metals in sludges had caused phytotoxic problems and plant losses, and he asked Dr Davis to comment on this.

Mr S. W. Bailey (ADAS/MAFF) said that disposal of spent sheep-dip was sometimes a serious problem, and there were some situations when it was very difficult to give acceptable advice on its disposal. He asked Mr Johnson and Mr Harris if they had treated spent sheep-dip with the Sentinel equipment?

Mr R. G. Toms (UK Centre for Economic and Environmental Development) said that much of the presentation had been concerned with the use of chemicals to grow products of good appearance and easy to sell. He said that there had been little mention of any uptake of pesticides ot even the taste of the produce. He was concerned that there had been a tendency to go ahead with new ways of farming without looking at the consequences. Surely the uptake of pesticides was just as important as producing a good-looking product. He asked what government research was doing about this?

Mr N. Whitley (States of Jersey Agricultural Department) asked, with regard to the Sentinel equipment, if there were limitations or whether it could be used for all chemicals? He asked the authors to enlarge on the total use that this equipment could be put to.

Mr N. C. Oxley (Howard Humphreys) was surprised that no mention had been made in either papers 1 or 2 of organic farming, and wondered to what extent organic production contributed to total agricultural production?

Mr J. Morris (Silsoe College) said that it seemed to him that we were very much prisoners of our beliefs and vested interests. The message that came across to him was very much one of continued use of agrochemicals, more use, and perhaps the development of new chemicals. He said it would be of more interest to identify alternatives to chemical use—perhaps the use of natural predators, and also changes in farm production methods. In many respects in the UK the move toward the use of more agrochemicalshad allowed a move to winter-oriented cropping and had allowed the sustainability of cropping systems that otherwise would not be feasible. In that respect there had been created a high degree of dependence on the chemical industry by the farmer. Reference had been made to the use of chemicals to control aflatoxin in cereals and oil seeds, but there were alternative solutions to this. One in particular was the use of good drying and storage methods. In those places where aflatoxin had been a problem, it had been more a case of growing crops in conditions which were not suited to those crops, and harvesting problems had been created as a result. Using chemical controls was treating symptoms, not causes.

Mr R. Allcock (Tay River Purification Board) said that he was interested in the Sentinel equipment because he had experienced many problems in his area in terms of disposal of pesticides and end-of-tank washings. He asked if users had been able to find local councils with suitable waste disposal licences who could remove the contaminated sludge produced?

Mr R. M. Walls (West Hampshire Water Company) said that his problem was analysing for the pesticides that may or may not be present in the potable water. This was an expensive consideration. He said the problem would be easier if one actually knew what pesticides were being used within a catchment area. This was something that appeared to have been resisted in the past, and he asked if Mr Malcolm could comment on this?

Mrs V. J. Hamlin-Wright (Stratton Streles Estates Ltd) asked for further comment on the Sentinel equipment. She asked what process of eligibility would be needed to qualify for the 50% grant?

With regard to the disposal of sewage sludge to land, she asked Dr Davis why only 40% of it was being utilized and whether the remainder could be used, or was it lack of available farming land?

Mr K. Wade (Welsh Water) said that one issue that had not been raised so far was the biodegradability of agrochemicals. He asked Mr Johnson and Mr Harris whether ICI were looking to produce a new generation of agrochemicals in the 1990s and beyond, where one would not have to treat a problem because the problem would disappear through biodegradability?

Dr H. F. Cook (Wye College) asked Dr Davis what might happen in the event of heavy metal levels exceeding the EC set levels? Were such sludges likely to be blended with other sources of sewage sludge or disposed of in another way, or in practice were levels likely to be controlled by adjustment of application rates to the soil?

Mr J. Turley (Yorkshire Water Authority) referred to Dr Davis's comment that it was a pity that more sewage sludge did not go to land, and that money was being wasted on incineration. He asked if it would be feasible to utilize incineration in large urbanized areas where there was a shortage of land within economic transport distance of works?

Mr H. A. Stevens (Boythorpe Ltd) said that with regard to the forthcoming legislation on slurry storage periods, did Dr Davis see less agricultural land becoming available on which to spread slurry?

Mr D. N. Harris (Anglian Water Authority) said he wished to confine his remarks to the Carbo-Flo plant for the reduction of pesticide levels in farm wastes to levels acceptable by water authorities for discharge to sewer, watercourse, or soakaway. The capital cost of £4500 for treating 1000 l aliquots of waste looked most attractive, but the operating costs needed to be considered, especially in replacing the activated carbon (40 m^3), which at current costs of approx. £550 per m^3 would amount to £22 000. Had the authors considered regenerating the exhausted activated carbon which, if combined with spent activated carbon produced from other similar plants, could reduce the cost to 50% of that of fresh activated carbon.

Mr B. Sinkins (South-West Water Authority) said that he was interested in the slide of sludge injection which showed almost a bowling green quality after injection. He said that in the South-West this had not been found to be the case, and he asked the author to comment on where they might be going wrong.

Mr B. D. Ogden (Yorkshire NRA) asked Mr Harris for comment in relation to the practical aspects of farmers using the Carbo-Flo equipment. He asked if someone could put forward a proposal for, say, a regional collection and treatment centre as had been done for other chemicals? He asked for the views of the commercial interests on that aspect and whether they could provide such a service?

He asked Dr Davis if he considered that the sludge disposal to land operation was 'passé' in view of the increasing environmental pressures, the new incinerator technology, and the pressures on trade effluent control which were down to limits of practical technology.

Mr S. Kirk (Yorkshire Water Authority) said that with regard to the identification of pesticides in groundwater, it was difficult to obtain data on the use of pesticides in a catchment. Studies were based on mass information which was based on extrapolations. Additionally, earlier studies by the WRc had pointed out that a large proportion of the pesticides found in groundwater were derived from non-agricultural sources, and he considered that this point should be stressed to balance the issue. Such sources were, for example, from roadsides and British Rail property where little reliable record of use was kept.

Mr J. R. Brook (Yorkshire NRA Unit) said that there seemed to be an increasing list of agrochemicals that had been produced and used. With some, problems had been found and they had subsequently been withdrawn. He asked the authors to comment on the likelihood of the chemicals currently in use eventually being withdrawn.

Mr A. R. Staniforth (Reading Agricultural Consultants), in a written contribution, said that Dr Davis had emphasized the importance of disposal of sewage to agricultural land and quoted the recent report of the USEPA which considered it to be the best of all methods of disposal. Dr Davis also briefly mentioned composting as a method of sewage treatment. At a recent French national symposium at Poitiers, a number of French sewage composting enterprises were described and demonstrated, and it was reported that 10% of US sewage was now being composted. It was worth mentioning that straw/sewage composting was being carried out by the Southern, Thames, and Anglian Water Authorities in this country on a pilot scale, and his firm were carrying out field trials with the product. The compost had a number of advantages, and that method of treatment and disposal could well be cheaper for the water authorities than other methods. It also had the advantage that another source of pollution, straw burning, could be dealt with at the same time.

Authors' replies to discussion on Papers 1, 2, and 3

PAPER 1

Mr J. Malcolm said that there had been several questions and comments with regard to the safety of pesticides, both in water and in food. He considered that it would be useful to go through the process that occurred before chemicals were accepted. Applications for new product approval had to go before the Advisory Committee on Pesticides, where the manufacturers had to present details of the chemicals and produce a large body of data to be examined. The Advisory Committee was a body of independent experts set up by the Ministry of Agriculture, Fisheries and Food, specifically charged with the job of looking at the chemicals to assess their safety as well as looking at questions of their efficacy. The process was very lengthy, and a good deal of care was taken about the safety of chemicals and a very wide safety margin was used.

He said that under the Food and Environment Protection Act which came into force during 1988, there were many rules about how the operator should use chemicals. They could be used only in accordance with the label instructions, and codes of practice were available to ensure that farmers sprayed properly.

There was a system within the EC for determining what were termed maximum residue levels of pesticide in food. The maximum residue level was not the highest amount of pesticide that a product might safely contain, but was used as an indicator of whether the pesticide had been properly applied. If the operator followed the rules of the Act and followed the manufacturer's instructions, then the maximum residue level should not be exceeded. Registration authorities always ensured that the maximum residue level was well within the safe level of exposure to any pesticide.

He said that in some cases where it was difficult to determine precisely the safety standards for the acceptable daily intakes of a pesticide, the acceptable daily intake was set as low as 0.0007 of what was the highest level of intake at which no effect was shown (the 'no observable effect' level). Hence, quite properly, there was a very wide safety margin applied in the registration of chemicals and in the setting of maximum residue levels.

Mr Malcolm was very critical of the simplistic approach of those who said that finding, say, an apple with twice the maximum pesticide residue level present meant that eating the apple would lead to poisoning. He described this as a totally irresponsible approach to take. Finding a product with twice the maximum residue level present might be an indication that the pesticide had been over-used, but it was not an indication that that would be harmful.

With regard to pesticides in water, Mr Malcolm said that under the EC Drinking Water Directive, the amount of pesticide that was allowed was one part per ten thousand million for any one pesticide, and five parts in ten thousand million for pesticides as a group. He said that the standard was set without any regard to toxicological evidence or consideration.

Mr Malcolm said the fact that a pesticide was measurable in water did not necessarily mean that it was harmful. However, having said that, it could be that there were some pesticides for which there should be a safety standard of below one part in ten thousand million. It was certainly the case that there were many pesticides which would be perfectly safe in drinking water at levels of more than the standard. He said that what was a great pity was that the EC had not chosen to use toxicological evidence or to set standards which were realistic in toxicological terms, so that water supply authorities and farmers and others could concentrate on seeking alternatives to any which were potentially harmful. He said that he had no doubt that it was quite proper for the chemical companies and for governments to be seeking to find more environmentally friendly pesticides, and the agrochemical companies were working on that.

Mr Malcolm said that it was necessary to look at some of the pesticides which were given approval a long time ago, which the Government was in fact now doing. The fact that approvals were given twenty or more years ago did not mean that those pesticides were now unsafe, but when those pesticides were tested for residues, etc, the measuring standards that were now available did not exist. It was now possible to measure in parts per thousand million, whereas twenty years ago it was difficult to measure in parts per million. It was therefore wise to re-assess some of those older pesticides to confirm their safety.

He said that the NFU supported Government action to withdraw any pesticides which are harmful to the operator, consumer, or the environment.

With regard to the organization of the collection of a chemical that had been withdrawn from use, it was possible to organize a reverse supply chain. In such a case, with one chemical with one chemical with known qualities in fairly readily assessed quantities, it was fairly easy. However, it would be much more difficult to operate where one would be collecting a mixture of containers of uncertain age, in uncertain amounts, and different chemicals. In fact, in East Anglia, the NFU was having a look at the possibility of operating a trial reverse chain to get off farms any old and unwanted pesticides and remove them safely.

He accepted that it was difficult if one was tracing something like 350 pesticides to know which ones to look for. Under the Food and Environment Protection Act, farmers were now obliged to keep detailed records of the fields and the crops on which they had sprayed the chemicals; of the extent to which they sprayed, the dates, etc. It should not be impossible to get together with the water suppliers to see if a reporting system could be set up, if that would be helpful.

He said that the NFU had suggested to the Government that there should be some trial water protection zones set up. This had been done with the problem of nitrates in mind, but it had been suggested that tacked onto that work there should also be monitoring of pesticide use and chemical use generally, and the effects of pesticides going into the watercourses. If Government agreement was obtained to set up such water protection zones, they would have to be looked at in co-operation with the water authorities, the DoE, the MAFF, the NFU, and others. More information would then be available about the extent of the problem.

He said that if the kind of comment that the representative from the Friends of the Earth had made about there being hundreds of old and dangerous pesticides were true, then people in the UK would have been 'dropping like flies'. This had not been the case because, by and large, there was little wrong with the water that people in the UK drank, and one expected that the water would continue to be perfectly drinkable. However, it was essential that we looked carefully and ensured that there was a wide safety margin.

Mr Malcolm said that integrated pest management was already practised in a number of agricultural areas, particularly in the glasshouse sector, where it was unusual nowadays to have a large number of pesticides because predators were released inside the glasshouse to attack pests such as red spider mite and whitefly.

He said that as far as organics were concerned, at the moment less than 1% of the UK food consumption was organically grown. Also, less than 1% of the production in the EC was organically grown. He said that the most optimistic forecasts put organic products at only 10% of the market by the year 2000. Hence the great majority of our food would continue to be grown using the benefits of agrochemicals.

PAPER 2

Mr D. A. Harris, in response to Mr Guiver, said that new molecules were not often developed just for a single market. They could not be developed for a market as small as the UK. They had to be truly international products and had to have application on crops of the World like maize or cotton in addition to the crops of temperate Europe. Development could take five years as a minimum, and in countries with a long registration period it could take considerably longer.

The cost could be up to about £40 million before the developing company got anything in return. Hence the investment was a very major one, and occasionally products were taken right through to the end of that process and then had to be abandoned.

With regard to the number of chemicals withdrawn from the market, he said that his feeling was that very few chemicals did get withdrawn after they had been marketed, because the procedure for their development was so onerous in the first place.

On the topic of the Carbo-Flo process and the Sentinel plant, Mr Harris said that the question of sheep-dip was one of considerable interest. Trials had been carried out in collaboration with the Yorkshire Water Authority, but the Allman Co., who were marketing the plant, had very small resources and been struggling to cope with the demand. More work was to be done, but it was considered that the active ingredients in sheep-dip could be handled by the process perfectly well. The problem was that sheep-dip had 10% lanolin in it, together with much wool and faeces, which were problems that had to be dealt with by pretreatment. It is the development of the pretreatment requirements in a handy and economic form which is being considered.

On the question of costs, Mr Harris said that the 50% grant on capital costs was undoubtedly going to be of great help in reducing the cost of an installation. The cost of the collecting tank and other parts of the installation might well be in excess of the Sentinel plant itself. In such case there was an opportunity for a contract service. He said that he did not know of one that was being operated at present, but contractors were giving it consideration. It was possible to consider sharing a plant between four or five farmers—either moving a plant physically from site to site to treat effluent collected in a system on each site, or using a road tanker to bring the effluent to a central place.

He said that the time cycle was about 3.5 hours. The dosing time for the chemicals into the top of the tank took just a few minutes. The agitator was run for about 10 minutes to stir it up. It was then switched off and settlement took place over 30–45 minutes. The liquid was then allowed to run through the filtration system, which took about 3 hours. It was therefore possible to treat two batches during a normal working day if necessary. It did not have to be attended for the whole time.

Mr Harris said that agrochemical companies were engaged in the development of alternative technologies. There were already insect pheromones which were used to bait insects in traps so that they did not have to be sprayed with a toxic material. There were materials like anti-feeding agents which did not kill insects directly but prevented them from feeding. He said that they also had a big input into plant-breeding in order to breed into plants the desirable characteristics including resistance to pests and diseases, and the transfer of desirable characteristics from one plant to another.

With reference to the required qualifications for the 50% grant, he said that the Agricultural Engineers' Association had recently issued a two-page note, and his interpretation was that the Farm and Conservation Grant Scheme would allow for all storage tanks and chambers, for treatment plant systems including permanent pipework, and fixed disposal facilities. Grants were not available for maintenance and repair nor for mobile plant and equipment such as mobile irrigation systems.

Mr Harris said that he had some experience over the years with injection equipment for slurry; the Paraplow being ICI's favourite loosening device. He said that there was a slurry injection machine which he believed was called the Western which used the Paraplow, and it could produce an almost bowling green finish.

Continuing the reply to the discussion on Paper 2, **Mr K. S. Johnson** said that the Sentinel equipment was devised for the dilute effluents produced on the farm. It was not intended to treat and dispose of the surplus chemicals which remained in the containers at the end of a spraying operation. The limitations, to a large extent, depended on the availability of adsorption capacity of activated carbon. On the chemical flocculation side, an abundance of solvent or oils could get through to the activated carbon. Over 90% of the adsorption capacity in activated carbon was inside the particle, so it was quite easy to see that if oil emerged into the carbon, it could coat the particles and cut off the adsorption sites, which would reduce the capacity of the carbon quite dramatically. That was one of the problems with sheep-dip where there was a lot of lanolin and grease. For the treatment of sheep-dip a larger pretreatment vessel would be needed, some way of removing the grease, and possibly prefiltration other than sand filtration, before the liquors could pass to the activated carbon. He said that he had no concern about the active ingredients being removed by the Sentinel equipment.

He said that the chemical flocculation stage had scope for use in the treatment of other effluents, being essentially a lime/iron system, but other flocculents could be used. He stressed that it would be important to consider each type of effluent on its merits, and he did not recommend any diversion from pesticides with the equipment without some prior investigation.

He said that his researches into the use of the activated-sludge and biological filter processes dated back to the early 1970s when the pesticides were in the main the more persistent types such as the chlorinated hydrocarbons. The approach had been to feed systematically, and increase clarified effluents in admixture with settled sewage. Up to 30% mixtures of clarified effluent had been achieved, and there had been no indications with acclimatized bacteria that the pesticides themselves had any undue adverse effect on the flora and fauna of the systems. However, there were concentrations of undegraded pesticides, such as lindane, emerging in the final effluents. For that reason it had been decided that those were not reliable methods for the treatment of pesticide effluent.

He said that there were a number of factors to consider with the use of activated carbon. There were many grades available. It was his his experience that the coconut shell carbons were not suitable for that application because of their very small pore size, which would not admit the larger molecular materials. It had been found that specially prepared coal-based carbons had the best effect overall.

The carbon cost about £1.50/kg, and in a commercial situation it was economic to have it regenerated, which was a way of disposing of the entrained pesticide on the carbon at very high temperatures. The losses were about 20% owing to burning and attrition which was made up with new carbon. However, in the field with a 25 kg pack in isolation it was not a candidate for regeneration because of the logistics of collecting them all together. He foresaw the emergence of replaceable modules, where the module which was exhausted could be returned to a central point and replaced with a new one.

Mr Johnson said that the amount of sludge produced in bulk terms was fairly small, about 100 kg of dried sludge in a season. Another aspect to remember was that the current trend and regime of pesticides was of lower persistence (weeks rather than years), and they were also prone to degradation by alkaline conditions. The sludges from the process were alkaline (pH 10–12). He said that the DoE Waste Management Paper (No. 21) recommended the allowance of 20 ppm of active pesticide for disposal in admixture with refuse. For example, a 500 t/d landfill site appropriately licensed could accept for disposal up to a tonne of sludge with a 1% active equivalent. He considered that it was more convenient to use the services of a reputable waste disposal contractor. A figure of about 50 p/kg collection price seemed to be about the order of cost.

PAPER 3

Dr Davis, in response to Mr Guiver, said that the £6 million for the value of nutrients in sewage sludge reflected the 40% of sludge that went to land at present. Theoretically one might assess the total value at £15 million but in reality that would not be attained because there were good reasons why the rest of the sludge did not go to agricultural land.

He said that anaerobic digestion certainly produced a sludge that was very suitable for using on the land, and it was probable that more sludge of that type would be made available because of the requirements of the EC Directive and the Code of Practice.

Stricter industrial effluent controls were limiting the quantities of contaminants that got into sewage and hence into sludge. He said that when it came to seeing phytotoxic effects it was quite difficult to achieve these in field trials, even when sludge was dosed at several times over the maximum permissible soil limits.

In answer to Mrs Hamlin-Wright, he said that some of the remainder of sludge that did not go to land was too contaminated, but the main reason was that there were cheaper disposal options, particularly for big conurbations where the sewage treatment works was a great distance from accessible agricultural land. The optimum travelling distance was less than 10 km.

He said that when soil metal levels were exceeded, sludge spreading would have to stop, but as sludges were getting cleaner one might hope that it would take applications repeated on the same piece of land for a period of 30–50 years or more before that stage was reached.

In response to Mr Turley, Dr Davis said that he did not wish to suggest that incineration was not an appropriate thing to do with sewage sludge. He thought that the important thing in sludge disposal was to have an array of disposal options available, and there was no doubt that in some circumstances incineration would be the best and the only sensible option to pursue, but he had no doubt that it was necessary to retain the agricultural outlet at least for the time being.

Dr Davis said that the Code of Practice required a storage period of three months as one of the treatment options. There were various other treatment processes where the amount of time that the sludge had to be held was much less. He thought that need not necessarily lead to difficulties in getting the sludge onto the land, although it was true to say that there seemed to be an increasingly shorter period when the land was free of a standing crop and available to spread sludge.

With regard to the comments about sludge injection performance, he said that he was pleased to say that the WRc ran a training course, in liason with Silsoe College, on sludge injection—he suggested that attendance at this might resolve the problems that Mr Sinkins had referred to. However, he confirmed that there were soil conditions where injection was simply not appropriate, and the WRc code of practice on injection did draw attention to the limitations of this method of application.

He said that he regretted that sludge to land might always be prone to the occasional 'shock horror headline' as were most sludge disposal outlets. However, it had been used in its present form for about 30 years and had been very seriously looked at from the point of view of environmental effects and was subject to careful control and monitoring. He thought that certainly for those sewage treatment works where agricultural land was readily accessible, and bearing in mind the attraction of recycling of nutrients and organic matter, the sludge to land policy was here to stay. Alternative outlets needed evaluation and there might, for instance, be an increase in incineration as some of the other sludge outlets became less accessible.

Chapter 2

Certain forms of pollution have been attributed to agricultural practices. This chapter groups together contributions from practitioners and regulators on the nature and extent of the problem. Groundwater pollution, particularly with nitrates and pesticides, are thought to pose potential risks to public health. Processes are highlighted which take place in aquifers to remove, alter, or attenuate contaminants. Changes in agricultural practices in the latter half of the twentieth century are believed to be a major contributing factor to the decline in river water quality. Remedies and initiatives are outlined and discussed from the point of view of the regulatory authorities.

Research projects have been identified to explore ways of reducing the impact of farm wastes on river quality. The encouraging outcome is the production of guidelines for catchment management leading to demonstrable improvement in river quality.

4

Contamination of groundwaters
from farming activities

H. G. Headworth[†]

ABSTRACT

There is wide concern in developed countries that groundwater resources are deteriorating in the long-term, both in quantity and quality. Some causes of pollution have a largely agricultural origin of which nitrates and pesticides are thought to pose serious risks to human health. This paper briefly reviews the sources of agricultural pollution to groundwater and the extent of the problem, and it summarizes the various processes which take place in aquifers to remove, alter, or attenuate contaminants.

KEY WORDS Contamination Groundwater Farming Pollution Nitrates Phosphates Pesticides

INTRODUCTION

There is wide concern in developed countries that groundwater resources are deteriorating in the long-term, both in quantity and quality. While groundwater depletion can be a serious regional problem, the enforcement of stronger controls can lead to the restoration of groundwater levels. This is not true of groundwater pollution which can be a national problem challenging technical understanding, economic restraint, and political will. Groundwater pollution is of major concern mainly because of the implications for human health when it is used for drinking water. Many of the most serious sources of health risk, such as bacterial and viral infection and toxic metals, do not have a specific agricultural origin. However, nitrates, as a cause of methaemoglobinaemia, and pesticides, with a suspected disturbingly wide array of health effects, have a largely agricultural origin, and they both pose threats to groundwater [1].

Concern over the problems caused by farming activities has grown apace in Europe in the last couple of years, and several countries, notably Germany, The Netherlands, and Denmark, are taking active steps to control nitrogen inputs into groundwater. As a result of their 1987 Soil Protection Act, the Dutch government has established groundwater

[†] Southern Water Authority, Guildbourne House, Chatsworth Road, Worthing, West Sussex.

quality reference values for their soils and groundwater for organic and inorganic contaminants, and has allocated substantial sums for the clean-up of contaminated soils.

Many countries now have aquifer protection policies backed by various degrees of statutory teeth. These APPs are generally based on travel-time protection zones which specify comprehensive controls over sources of point pollution. However, they are largely impotent against diffuse, region-wide sources of pollution, and deriving suitable controls is posing political, and not just technical, problems.

This paper briefly reviews the sources of agricultural pollution to groundwater and the extent of the problem, concentrating mainly on diffuse sources. But first it looks at the various processes which take place in aquifers which remove, alter, or attenuate contaminants.

HYDROGEOLOGICAL PROCESSES

By virtue of their composition, fabric, and thickness, aquifers have the capacity to greatly affect the passage of contaminants through them [2]. The attenuation of contaminants in groundwater results from a variety of physical, geochemical, and biochemical processes. The two most important physical processes are hydrodynamic dispersion and filtration. The importance of each of these processes is dependent on the size and nature of the contaminant present. If bacteria are present, then filtration will probably be an effective method of removal provided that clays and silts are present in the porous media. Dissolved solids, in ionic or molecular form, are likely to be strongly affected by hydrodynamic dispersion, which is the spreading of such ions and molecules within the pores and fractures by the groundwater velocity.

Of the various geochemical processes only adsorption, precipitation, and certain microbial reactions can lead to the actual removal of contaminants from groundwater. Other geochemical processes of complexation and acid–base reactions affect groundwater in various ways which may or may not enhance such removal of contaminants.

The removal of contaminants from groundwater by adsorption and precipitation may be only a temporary phenomenon since desorption and dissolution may subsequently release the contaminants back to the groundwater once the contaminant front has passed and more dilute waters take its place. Adsorption by ion exchange reduces the natural concentrations of contaminants and retards their movement without reducing the total amounts released. This allows certain decay processes, such as biodegradation, to further reduce the contaminant concentrations in groundwater since more time is available for such processes to occur.

Biochemical processes are very important in contaminant hydrogeology. Bacteria are unicellular microbes with rigid cell walls. They are roughly the size of small silt and large clay particles. Microbes, which cause the decomposition of organic wastes, require complex organic molecules, such as carbohydrates, as nutrients. Consequently they use carbon, hydrogen, oxygen, phosphorus, nitrogen, and sulphur to thrive, provided that suitable temperature and pH conditions exist. While most bacteria exhibit optimum growth rates at 30–40°C, some grow well at near zero temperatures. Some families of bacteria have adapted to pH environments outside the optimum range of 6.5 to 7.5.

Biochemical processes initiated by bacteria may have beneficial or detrimental effects on contaminants, depending on the type of organism present and the conditions in which they exist. Beneficial effects include the purification of contaminated water in aerobic conditions in which organic contaminants are broken down to relatively innocuous products such as carbon dioxide, water, nitrate, and sulphate, and the uptake of nitrogen, sulphur, carbon, and phosphorus, which may be removed from groundwater as a consequence. Detrimental effects of microbial activity can occur with the consumption of all existing free oxygen, producing methane, hydrogen, ammonia, and hydrogen sulphide. Furthermore, in such anaerobic environments various metals may become relatively more soluble (for example iron and manganese).

POINT SOURCES OF AGRICULTURAL POLLUTION

Groundwater is vulnerable to a wide range of point source pollution from agriculture, but actual quoted instances are not very common. In the survey *Water pollution from farm wastes in England and Wales* [3], in which 990 serious farm pollution incidents are recorded, reference to only one case of borehole pollution is made, in Wessex. Manure heaps, farmyard drainage, disposal of waste and surplus biocides, spent sheep dips, and silage liquors all pose threats of which the last is the most serious. Aldrick, in a personal communication, notes numerous cases in Yorkshire where springs in karstic limestone have been polluted from silage clamps. Yorkshire Water Authority are also concerned about the practice of disposing of sheep dip liquors to soakaways in the Chalk, and they have established a multi-disciplinary groups to investigate this and allied problems.

Pollution from farm wastes is the subject of two papers at this Symposium, and I do not intend to dwell further on this topic in this paper.

PESTICIDES

Groundwater contamination from pesticides is causing growing anxiety. Many believe it will become a major environmental issue in the 1990s. (Pesticide is here used as a general expression to include herbicides, insecticides, and fungicides, unless these are referred to specifically.)

Over recent years there has been a rise in the number of reported incidents of contamination of groundwater from organic chemicals [4]. This has resulted from the increasing use of more intensive agricultural methods, and the development of more advanced analytical instrumentation. Little is known about the long-term toxicity of many of these chemicals. Pesticides, such as triazines, are used extensively for non-agricultural use, which clearly are giving rise to the high concentrations found.

The occurence of pesticides in groundwater is widely recorded. Milde *et al.* [5], in a literature review, reports on the occurrence of 22 different pesticides in 34 localities in the USA and Europe. Atrazine is the most widely reported, with concentrations up to 7 µg/l. In Britain the most extensive survey of pesticides in surface waters and groundwaters has taken place in the Anglian region where the herbicide usage on cereals has been the greatest. Croll [6] reports that while ten different pesticides are found frequently in surface waters, they are less frequently detected in underground waters at concentrations much greater than 0.1 µg/l. On the basis of 200 samples of groundwater

Croll tabulates the following occurrences (see Table 4.1):

Table 4.1. Pesticides detected in groundwaters (after Croll, 1988)

Pesticide	Concentration range, µg/l	Occurrence, % of samples
Mecoprop	0.1 to 0.38	3
MCPA	0.12	0.5
Atrazine	0.02 to 0.43	28
Simazine	0.02 to 0.26	9
2, 4 – D	0.11 to 0.2	1

Knowledge of the occurrence of pesticides in groundwater in the rest of England and Wales is patchy, as is the monitoring being carried out. Milde states that the main factors influencing the contamination potential of any pesticide are molecular structure, formulation, metabolism, water solubility, adsorption by soil, the retardation factor, and resistence to chemical and biochemical degradation. The time, quantity, and frequency of application, as well as the local hydrogeological and meteorological conditions, will all affect the quantity of any pesticides reaching the water table.

A study carried out by the Water Research Centre recorded Steenvoorden's [7] estimated relative solubilities and rates of degradation of certain pesticides (see Table 4.2):

Table 4.2. General solubility, mobility, and rate of degradation of some pesticides (after Steenvoorden, 1976)

Pesticide group	Example	Mobility	Solubility	Rate of degradation
Organo-chlorine	Lindane	very low	very low	years
Organo-phosphorus	parathion	low	low	weeks to months
Uracils	bromocil	low–high	low–high	months to years
Phenoxyalkanoic acids	MCPA	high	high	weeks to months
Triazines	atrazine	low	low	weeks to months
Carbamates	barban	low	low–high	days to weeks

Steenvoorden concluded that the organo-chlorine pesticides are long-lived, have a lower solubility, and exhibit chronic toxicity to all life-forms. They are strongly adsorbed by soil and are unlikely to be found at depth. The organo-phosphorus compounds exhibit a wide range of solubilities but have a low mobility, equal to the organo-chlorines. They do, however, degrade more rapidly with a persistence of 0.1% after 15 years, compared with 3–20% for the organo-chlorines.

The uracil group, which comprises many herbicides, have high solubility and are rather persistent. In general they are fairly mobile, and some are toxic. Bromocil is a good example of a mobile uracil. Other groups of biocides, such as triazines and carbamates, can differ in solubility and mobility, but the risk of groundwater pollution should be less because of their higher breakdown rate. The large number of occurrences of atrazines in groundwater suggests wide usage for non-agricultural purposes.

A study is under way by WRc [8] on pesticides in the Granta catchment in Anglia. As part of its objectives this study is looking at the transport and fate of pesticides in aquifers, as well as methods of prediction, and it should prove invaluable in giving greater understanding of the long-term pollution risk from pesticides.

NITRATES

The topic of rising nitrate concentrations in groundwater is now a well-trodden theme which has been the subject of extensive research and international debate. Some argument still surrounds the cause of rising nitrate levels and the confidence in the long-term predictions, but these issues have now largely abated. In England and Wales 125 groundwater sources supplying 1.8 m people had nitrate concentrations in 1983 or 1984 exceeding 50 mg/l (throughout this paper expressed as nitrate rather than nitrogen) [9]. 105 of these sources exceeded 50 mg/l in 1984 compared with 90 groundwater sources in 1980 and 60 sources in 1970. Areas in Anglian and the Midlands are worst affected where nitrates are rising in about 100 sources. The most severely contaminated group of sources are still those in the Jurassic Limestone of central Lincolnshire, where some boreholes have had to be taken out of supply, used intermittently, or used only when blended with low-nitrate water.

Research undertaken

The threat of rising nitrates in groundwaters was first brought to public attention by Foster & Crease [10] in a study carried out in East Yorkshire. They reported on sharp rises in nitrate concentrations occurring in 1969/71 in certain Chalk sources. They drew attention to the apparent anomaly between the traditional understanding of speedy transit of infiltrating water through the unsaturated Chalk aquifer down to the water table (1 m/d) and the environmental tritium studies which showed very slow downward rates of movement (1 m/y).

Foster [11] postulated that while relatively rapid unsaturated groundwater flow could occur in fissures, the downward movement of tritium (or other solutes) was largely retarded as a result of ion exchange by molecular diffusion between the mobile fissure water and the static pore water. It was subsequently shown mathematically that this mechanism can produce vertical distribution of solutes [12]. Diffusion operates as an exchange of solute between that contained in the water moving down through fissures and that contained in the pore water. If the concentration of solute in the fissure water is higher than in the pore water, then the solute diffuses from the fissures into the pore water, and if the fissure water has a lower solute concentration than the pore water, then the reverse is the case. This process of exchange is dynamic and leads to a slowly moving downward front of solute (for example nitrate, chloride, sulphate) through the unsaturated aquifer zone. The rate of downward movement is influenced by the infiltration and the pore water pososity. It therefore varies with aquifer type and regional locality, and amounts to over 2 m/y in the case of the Sherwood Sandstone of the Midlands to 0.5 m/y in the Chalk of the Isle of Thanet in Kent [13]. Research undertaken by the British Geological Survey and the Water Research Centre since then has shown an indisputable link between farming practice and nitrates in aquifers. Below unfertilized and lightly fertilized permanent grassland nitrate concentrations in the unsaturated zone pore water are normally less than 10 mg/l. Below arable land, subject to large fertilizer application rates, nitrates can be well in excess of 50 mg/l with peaks exceeding 200 mg/l. The cultivation of unfertilized grassland can lead to the mineralization of organic nitrogen and can produce sharp nitrate peaks in the unsaturated column. High nitrate concentrations are accompanied by elevated levels of other solutes, notably sulphate, chloride, and some trace elements derived directly from fertilizers or mobilized from soils after ploughing.

Mathematical models developed and refined during the last ten years [12], [13], [14], [15] have led to a better understanding of the soil leaching rates and the diffusion process. They have enabled long-term predictions to be made of future nitrate trends. Groundwaters in the Midlands, East Anglia, Yorkshire, and North Kent are those most vulnerable to nitrate pollution. These are areas of low rainfall where arable and horticultural farming predominates. Depending on the dominance of fissure flow in the unsaturated zone and the depth to the water table, some aquifers, such as the Lincolnshire Limestone, respond quickly to nitrate leaching losses (5 to 10 years), while others, such as the Chalk of Thanet, are very slow to respond (30 to 40 years).

Recent studies

The implications to the water industry of these predictions were discussed at length in the 1986 Nitrates in Water Report, and several proposals were put forward to reduce leaching losses. These centred around limiting the rate or amount of nitrogen fertilizer applied, prohibiting the application of nitrogen fertilizer during certain periods, prohibiting some crops, or limiting land use to permanent grass or specific crops, and preventing the ploughing of grassland. The Report discussed the comparative economics of these agricultural controls and the treatment of high nitrate water by water undertakings. It also recommended that consideration be given to the use of protection zones around groundwater sources to control farming practice so as to reduce nitrate leaching losses from the soil.

After the publication of this Report, Severn–Trent undertook a detailed study in their Hatton catchment in conjunction with DoE and MAFF [16]. This compared the options for water treatment and the use of protection zones around sources. The Report concluded that for Hatton it would be relatively cheap to provide blending and treatment compared with other areas. Despite this, these costs were not significantly less than combined solutions with some land use controls. Elsewhere, where these favourable engineering factors might not be present, options requiring a modification of land use might offer clearer financial arrangements.

In 1988 DoE led a desk study [17] in collaboration with MAFF and five water undertakings which compared the cost to the water industry of blending, treatment, and source replacement with that to the farming industry of land use controls. In three of the ten catchments studied the response time to rising nitrates is so slow that no change in farming practice, even if effected now, would lead to nitrates declining to below their statutory limit. In eight areas, the preferred water option is less costly than the preferred agricultural option on a local cost basis, but on a national resource basis, having regard to the need to reduce food production, the preferred agricultural option is likely to be cheaper than the water option in seven areas.

In all but one area where an agricultural option was chosen, significant changes to farming practice would be needed. Agricultural options are likely to compare favourably with water options in areas of mixed farming underlain by speedy-response aquifers, where no cheap source of blending water exists and where rainfall is moderately high.

CONCLUSIONS:

(1) To maintain a high level of agricultural productivity and to continue to provide the high quality, readily available produce which the retailer and consumer demand, farmers in the UK will continue to rely on agrochemicals. However, it must be recognized that there is a continuing groundswell of opinion which voices legitimate concern about the long-term effects of the country's dependence on agrochemicals.

(2) As time progresses, the environmental and natural food lobbies are likely to strengthen, and with them restrictions on the farmer's freedom to use chemicals.

(3) The Food and Environment Protection Act, and the regulations which it has introduced into the use of pesticides, will help to improve the safe use of pesticides.

It will remain to be seen whether Ministers grasp the nettle of adopting a policy of selective source protection zones as a means of reducing leaching losses from nitrogen fertilizers.

REFERENCES

[1] Blake, J. (1987) Improved protection of water resources from long-term and cumulative pollution. *Paper to OECD Environment Committee Group on Natural Resource Management. December*, Paris.

[2] Jackson, R. E. (1980) Aquifer contamination and protection. *UNESCO Project 8.3 of the International Hydrological Programme.* Report of Project Working Group.

[3] Water Authorities Association (1987) *Water pollution from farm waste in England and Wales.* Report.

[4] Baxter, K. M. (1986) Organic loading to groundwater: A review. II Use of pesticides in England and Wales. *WRc Report 1198–M.*

[5] Milde, G., Pribyl, J., Kiper, M. & Friesel, P. (1982) Problems of pesticide use and the impact on groundwater. In: *IAH International Symposium Impact of Agricultural Activities on Groundwater*, Prague.

[6] Croll, B. T. (1988) Pesticides and other organic chemicals. In: *Symposium—Catchment Quality Control*, IWEM.

[7] Steenvoorden, J. H. A. M. (1976) Nitrogen, phosphate and biocides in groundwater as influenced by soil factors and agriculture. In: *Proceedings of Technical Meeting on Groundwater Pollution*, The Hague.

[8] Hennings, S. & Morgan, D. (1987) The Granta Catchment Pesticides Study: Progress Report. *WRc Report 1716–M/1.*

[9] Department of the Environment (1986) Nitrate in water—Report by the Nitrate Co-ordination Group. *Pollution Paper No. 26.*

[10] Foster, S. S. D. & Crease, R. L. (1974) Nitrate pollution of Chalk groundwater in East Yorkshire—a hydrogeological appraisal. *Jour. Inst. Wat. Eng.*, **28**, 178–194.

[11] Foster, S. S. D. (1975) The Chalk groundwater tritium anomaly—a possible explanation. *Jour. Hydrol.*, **25**, 159–165.

[12] Oakes, D. B. (1977) The movement of water and solutes through the unsaturated

zone of the Chalk of the United Kingdom. In: *Proc. 3rd Int. Hydrol. Symp.*, Colo. State Univ. Fort Collins, Colo.

[13] Southern Water (1985) Report on Thanet Nitrate Investigations. Internal report.

[14] Young, C. P., Hall, E. S. & Oakes, D. B. (1976) Nitrate in groundwater—studies on the Chalk near Winchester, Hampshire. *WRc Report*.

[15] Foster, S. S. D., Bridge, L. R., Geake, A. K., Lawrence, A. R. & Parker, J. M. (1986) The groundwater nitrate problem. *BGS Hydrogeological Report 86/2*.

[16] Severn–Trent Water (1988) The Hatton catchment nitrate study, *A report of a joint investigation on the control of nitrate in water supplies*.

[17] Department of the Environment (1988) The nitrate issue. A study of the economic and other consequences of various local options for limiting nitrate concentrations in drinking water. (In preparation).

5

Farm waste pollution

L. Beck (Fellow)[†]

ABSTRACT

This paper looks at post-war changes in farming and sees the latter half of the 20th century as suffering from the agricultural equivalent of the original industrial revolution; as we struggle to clean up the legacy of that first revolution in the dirty rivers, our once-clean rivers are now declining in quality, so that the net effect is one of running hard to stand still. Recent pollution statistics are reviewed, with some stories behind them. Remedies and initiatives are discussed from the point of view of the regulatory authorities.

KEY WORDS   Post-war history   Legal framework   Pollution statistics   Law enforcement   Treatment options   Political action

INTRODUCTION

Agricultural pollution has recently been much in the public eye, and the news media have focused on it in such a way that other forms of pollution have sometimes been partly eclipsed.

The problem of farm pollution has certainly grown, and we should examine the reasons, the attempts being made to solve the problem, and the likelihood of success.

BACKGROUND

Instructions on a healthy environment goes at least as far back as the Book of Deuteronomy

> '...it shall be, when thou wilt ease thyself...thou shalt...and shalt turn back and cover that which cometh from thee.'

> Deut. 23:13 (AV)

As with domestic waste, the waste from food production has, through the centuries, been absorbed and recycled by nature without creating serious imbalance. Farming, all

---

[†] Co-ordinating Manager (River Quality), Yorkshire Water.

the way from the goats and chickens in the backyards of early homesteads right up to the mid-twentieth century farms, was not intensive, they used little water, no sophisticated chemicals, and were widely scattered. Any water pollution was local, and the 'dilute-and-disperse' principle was in operation before anyone invented the phrase.

The post-war years

During the immediate post-war years the traditional small farm caused little significant pollution, and national attention given to agriculture was mainly in the context of increased food production.

The Rivers (Prevention of Pollution) Act of 1951 hardly scratched the surface of sewage and industrial effluent control since it dealt only with the control of new discharges. However, it introduced the offence of causing or knowingly permitting poisonous, noxious, or polluting matter to enter a watercourse, which has remained intact in current legislation, and has been used frequently in dealing with irregular discharges from farms.

It was not until the Rivers (Prevention of Pollution) Act of 1961 that we could face up to controlling existing effluents (including regular farm effluents). There were thousands of applications, and many of them related to farm house septic tank effluents, run-off from manure middens, collecting yards, etc. discharging to insignificant ditches. In Yorkshire, only those causing gross pollution and public complaint were dealt with, usually by consent refusals, to take effect after a negotiated period. This still left thousands of unresolved applications gathering dust between 1963 and the present day. (If these discharges still exist, they now have 'deemed consent' under COPA.)

Some authorities tried to deal with all their applications, either by Consent or Refusal, with various degrees of success. During this period (in which by the way the normal standard for almost any effluent was the 'Royal Commission' or '30/20' standard) authorities wrestled with ways of granting consents to insignificant farm effluents, but were never going to get within light years of a Royal Commission standard.

The Thames Conservancy, which had existed since 1857 and was well into the 'Royal Commission habit' before other River Boards were born, dealt with many by stretching the definition of the effluent to mean the receiving stream at the point where it left the farm boundary; it was here that the consent standards would be applied.

Others granted consents and applied standards direct to the effluents, knowing they would never be met. They then forgot about them, content in the knowledge that if somebody complained about pollution then the consent could be dug out, dusted off, and used as ammunition.

This somewhat tame approach continued through the 1960s, running parallel to a national attitude that farmers could do no wrong. The agricultural industry grew and intensified.

The progression of farm pollution from being a minor irritant to a subject of national concern can be traced through river board and water authority annual reports.

The Yorkshire Ouse River Board's Report for 1960/61 [9] contains not a single agricultural pollution statistic. (From a region with about 10% of England's dairy and beef herds and 20% of England's pigs) [10]. It records only 167 pollution incidents of any kind as being worthy of note (61 causing complaint), and they were associated with coal, coke, and gas, textiles, foodstuffs, leather, paper, sand and gravel, and miscellaneous. The 30 'miscellaneous' might have included some farm pollution, but this is not clear.

Some 36 000 words of commentary end with twelve lines to the effect that there has been 'several cases of pollution by piggery refuse and silage drainage' (and it is interesting to note that some of these farmers claims that their faulty drainage arrangements had been 'approved by the local Agricultural Executive'!).

This pattern continues to March 1964 [11] at which point a small section headed 'Farm effluents, etc.' appears, though there are still no farm pollution statistics.

By the first Report of the new Yorkshire Ouse & Hull River Authority (March 1966) [12] there are still no agricultural pollution statistics.

By March 1966 [12] FARMS appeared for the first time as a separate category of trade premises causing significant pollution incidents (13% of the total). By the following year and in all subsequent years, Annual Reports feature agricultural pollution as a regular and increasingly important subject.

And so in Yorkshire, in environmental impact terms, the agricultural equivalent of the Industrial Revolution materialized around the mid-1960s. This appeared to be typical of the country as a whole, judging by the spate of papers on this subject in the IWPC Journals, the highest concentration of which appeared in the period 1966–1978.

Since the formation of the water authorities in 1974, agricultural pollution statistics have continued to feature in annual reports, and in Yorkshire the trend has continued upwards (Table 5.1) [13], with a similar picture in England and Wales (Table 5.2) [8, 21].

Table 5.1. Pollution incidents in Yorkshire and N. Humberside

Year	Total incidents	Agricultural incidents	Agricultural share (%)
1976	1033	139	13.5
1977	854	106	12.4
1978	861	82	9.5
1979	968	117	12.1
1980	1071	156	14.6
1981	1136	149	13.1
1982	1020	156	15.3
1983	1165	194	16.6
1984	1536	227	14.8
1985	2006	328	16.4
1986	2166	246	11.4
1987	2323	336	14.5
1988 (Est)	2900	330	11.4

In Scotland, the Clyde River Purification Board reported 23% of all incidents in 1987 as being agricultural [14], noting particularly in the river Ayr catchment that farm pollution incidents had been on the increase over the previous three years. The Solway River Purification Board reported 59% of all incidents in 1987 as agricultural [15] with special reference to a 50% increase in silage incidents.

THE NATURE OF THE BEAST

Intensive animal farming is good at filling supermarket shelves and satisfying the 'needs' of the modern European, but it is less good at harmonizing with nature.

Table 5.2. Pollution incidents in England and Wales

Year	Total incidents	Agricultural incidents	Agricultural share (%)
1980	12500	1671	13.4
1981	12500	2367	18.9
1982	12300	2428	19.7
1983	15250	2795	18.3
1984	18648	2828	15.2
1985	19892	3510	17.6
1986	19892	3427	17.2
1987	23253	3890	16.7

Pigs, which used to be in small units, housed on straw, and allowed to root about in a field, are now housed in their thousands, and their waste output of about 4 l/hd d [1, 2, 3] is often boosted to three times that amount by the use of a liquid diet [4].

Dairy herds, which used to stand in straw-bedded housing when not grazing, are now in units of up to several hundred, and the twice-daily cleaning of the milking parlour, collecting yard, and dairy can produce up to 86 l/hd d of waste [5].

Hay making has given way to silage, which itself produces the most polluting effluent on the farm.

Arable farming is equally intensive, made possible by the use of chemicals. These chemicals are the cause of pollution at their point of use, their point of manufacture (the Sandoz incident) [6], and their point of storage (the Woodkirk incident) [7].

Another contributor to increased arable output is the use of artificial fertilizers which again are a potential source of pollution during manufacture, storage, and use, especially the liquids. The gradual build-up of nitrate in groundwaters is causing concern.

There has been a continuing climb in the size of dairy herds and piggeries (Fig. 5.1), the tonnage of silage (Fig. 5.2), and the moisture content of silage made (Fig. 5.3) [8].

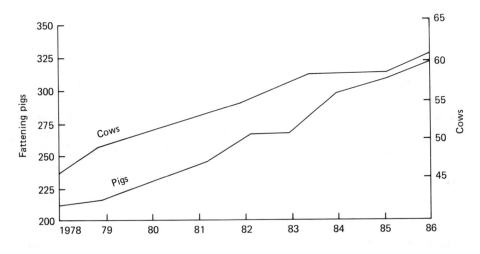

Fig. 5.1. Average herd size [8].

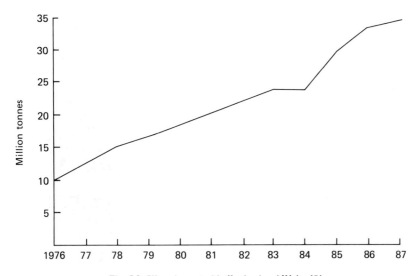

Fig. 5.2. Silage harvested in England and Wales [8].

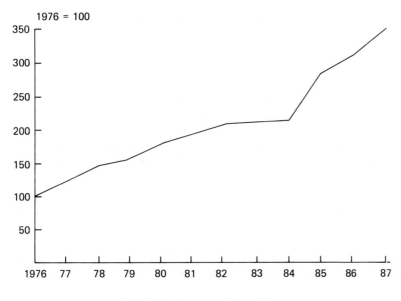

Fig. 5.3. Liquid content of silage [8].

Point source pollution arises from inadequate collection, storage, or disposal of animal slurries and silage effluents, and from the careless handling of agrochemicals.

Animal slurries and silage effluents are the most common causes of farm pollution incidents.

The relative impact of farm wastes on the environment is high (Fig. 5.4) [8], and is enhanced by the fact that incidents usually take place on so-called 'clean' river systems.

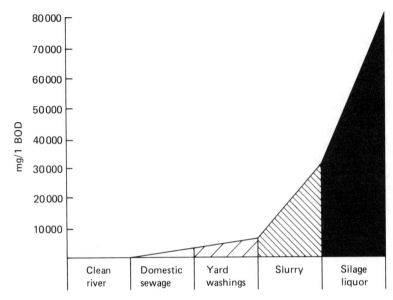

Fig. 5.4. Comparative strength of wastes [8].

ENVIRONMENTAL IMPACT

The nature of farm wastes and their effects is already well documented. It suffices to summarize here the main inputs and their general effects.

Animal wastes
BOD 10 000–36 000 mg/l [5]; ammonia levels which are toxic to fish and interfere with potable water treatment; decomposing solids which blanket the stream bed.

Vegetable processing wastes
BOD 80–12 000 mg/l depending on the vegetable type and the degree of processing [3].

Silage
Acidic; extraordinarily high BOD 30 000–80 000 mg/l [8]; promotes prolific growths of 'sewage fungus'. A small trickle has a devastating effect. (This year (1989), 38% of farm pollution incidents in Yorkshire and North Humberside have been connected with silage.)

Fertilizers
Ammonia levels which are toxic to fish and interfere with potable water treatment; elevated levels of nitrogen and phosphorus causing eutrophic conditions; algal taste problems in water supplies, and potentially high levels of nitrate in public supply.

Pesticides and herbicides
These are designed to kill, so it is no surprise that in the aquatic environment they destroy

flora and fauna and threaten the human population through the use of ground and surface waters for potable supply. (In 1988 Friends of the Earth drew attention to 300 UK potable supply sources with levels of pesticides in excess of the EC Drinking Water Directive. The most common materials were the herbicides Atrazine and Simazine.)

TREATMENT AND DISPOSAL OPTIONS

Animal wastes

Since the mid-1960s much work has been done on methods of treating farm animal wastes, including anaerobic lagoons, anaerobic digesters, aerobic lagoons, oxidation ditches, extended aeration units, barrier ditches, rotating biological contactors, biological filters, separators, composting, incineration, wet-air oxidation, drying, and various combination of all or some of these techniques.

The results, for the most part, have been hugely disappointing.

Anaerobic digestion produces gas, and thus has appeal as something which ought to have an economic return. There have been a number of experimental units, but they have proved expensive and problematical. Hobson [16] states 'to be successful, digesters have to be made relatively cheaply yet still capable of operating with minimum supervision from farm staff whose main job is away from the digester and who cannot call on the engineering and other back-up of the sewage works or the factory'. The object of the exercise is not to discharge an effluent to stream, but to produce energy and a slurry with reduced pathogens and reduced odour for land disposal.

Aerobic systems have been extensively tried. Some aerobic systems are mainly for odour control, such as the Plunging Jet Aerator [17]. In the Netherlands, in particular, where livestock numbers per hectare are four times those in the UK [27], work on the use of oxidation ditches has been well documented. A unit treating the waste from about 11 000 pigs dealt with 200 kg BOD/d and achieved a 96% reduction in BOD and a 67% reduction in ammonia [18]. Unfortunately, this still translated into an effluent with a BOD of 80 mg/l, and 400 mg/l ammonia N. Scheltinga's work produced oxidation ditch effluent BODs of between 10 and 100 mg/l, and ammonia N between 5 and 400 mg/l [19].

Work by the Water Research Centre in 1975 on aerobic systems led to no hopeful conclusions about the ability to produce a dischargeable effluent. Indeed, the point was made that final land disposal should be considered as a high a priority in farm waste treatment [5].

Barrier ditches have often been advanced as a useful system, but practical experience has usually encouraged a more pessimistic view. Willetts & Weller [4] describe this system, but warn that it is probably better to use the effluent for organic irrigation.

At a large well-managed ICI dairy farm in Somerset, a 10-section barrier ditch system was originally used for 'brown water' wastes, and it was reported at an Open Day in 1988 to have achieved a 97% reduction in BOD and solids. However, this still left an effluent BOD of about 80 mg/l and 155 mg/l solids. The system was modified so that 'brown water' went to storage, solids separator, and low-rate irrigation system, and the barrier ditch was used only to receive storm overflows from this system. In August 1988 samples taken by the farm and analysed by South West Water showed that under moderate weather conditions the BOD was reduced from 323 to 19 mg/l, ammonia N from 261 to 84 mg/l, and solids from 603 to 54 mg/l. However, this kind of performance was not maintained when subjected to storm conditions, with the effluent climbing back into three figures.

So far as poultry wastes are concerned, Riley [20] concluded that traditional methods of treatment did not appear suitable.

The general message for the average farmer has to be:

(1)   Forget about treatment.
(2)   Separate clean water from dirty water.
(3)   Contain slurries, manures, silage effluent.
(4)   Provide storage capacity to cater for weather and cropping.
(5)   Spread on land away from streams and in the right weather.

With regard to (4) Mason [26] has indicated a minimum of 4 months. Professor O'Callaghan (Newcastle University) said at the 1988 FBA Farm Waste Conference that 5 months was preferable.

With regard to (5) it is important not to exceed the crop nutrient requirement. Nielson [27] has suggested that for most crops, pig and cattle manure at 6–8% dry matter should not be applied at more than 50 m^3/ha, and poultry manure at 20% dry matter, 25 m^3/ha.

Vegetable processing waste
These can be treated by conventional biological processes provided that the load is carefully controlled and nutrient imbalance is corrected, to produce a 30:20 effluent or better.

Silage effluent
These are the alternatives:

(1)   Prevent effluent production by wilting the crop (Fig. 5.5) [22].
(2)   Be prepared for large volumes and have the necessary systems of collection, storage, and disposal.

Option (1) has always been enthusiastically promoted by the regulatory authorities, but they must now acknowledge that this particular battle has been lost: a combination of British weather and the current practice of several cuts per year with rapid harvesting guarantees an effluent problem (Fig. 5.6) [22].

Therefore the problems of option (2) have to be addressed. Proprietary absorbents can be added (which also enhance the feed value), but they will not remove the need for effluent storage.

Silo floors must be water-tight and constructed either by using concrete preferably with a water/cement ratio not greater than 0.4:1 (Fig. 5.7) [23] or hot-rolled asphalt. Internal and external effluent drains must be provided to direct effluent to a water-tight storage tank of large enough capacity to permit emptying as required without upsetting the farm routine. A special prefabricated acid-resistant tank is preferable.

Effluent can be spread on land after dilution with water or slurry. (NB It must NOT be mixed with slurry stored WITHIN BUILDINGS because of the danger of toxic gases.)

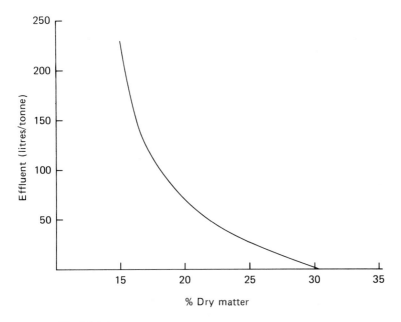

Fig. 5.5. Effects of dry matter content on silage effluent volume [22].

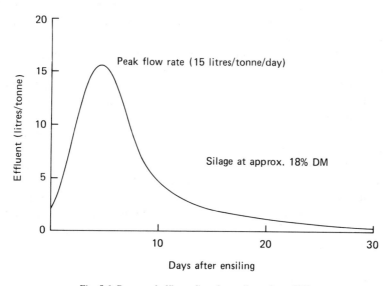

Fig. 5.6. Pattern of effluent flow from silage clamp [22].

In an interesting method now being used on many farms, the silage liquor is fed back to the cows. Henley Manor Farm in Somerset (ICI) practises this successfully, with all but a few days of peak effluent production being recycled in this manner, at a rate of consumption of about 10 l/hd d.

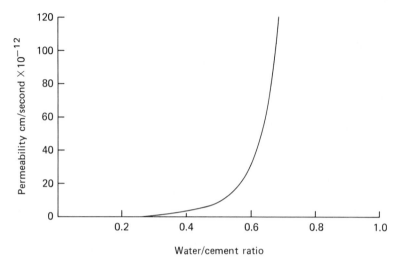

Fig. 5.7. Effect of water/cement ratio on permeability [26].

Pesticides

It is essential that pesticides be securely locked away in a purpose built store, built with fire-resistant materials, and having capacity for retaining spillages and/or fire fighting waters. Important advice to the farmer on storage is given by the HSE [28] and MAFF [29], and on usage by MAFF [30]. Guidance is given to officers of water and fire authorities on inspection of agrochemical stores for the British Agrochemical Standards Inspection Scheme (BASIS) in a document drawn up jointly by the Water Authorities Association, the Chief & Assistant Chief Fire Officers Association, and BASIS [31].

Discussion within the water industry is going on concerning disposal of end-of-spray residues and tank washings. The temptation to empty into nearby ditches, or farm drains, has to be rigorously resisted. If such residues cannot be re-used in the next mix, they should be spread over land at normal distribution rates.

The use of soakaways is not a practice to be generally encouraged. However, in certain cases such a disposal route might be permitted if there is no danger of breaching the EC Groundwater Directive or of affecting water supplies. Current legislation requires that the consent of the water pollution control authority be obtained before discharging this trade effluent into land. It has to be stressed that critical levels of these substances are measured in tenths of a microgram per litre.

WHAT'S HAPPENING DOWN ON THE FARM?

Here are the factors which have contributed in the past two years to the rising tide of farm pollution incidents [8]:

○   Poor quality farm building work (DIY and contractors)
○   'Unexpected' weather

○ Land run-off from   – slurry overdosing
                                        – slurry spreading on frozen ground
                                        – slurry spreading during heavy rain
                                        – unmoved rainguns
                                        – irrigation pipes with leaking joints
○ Exclusive reliance on neighbours' land for disposal
○ Unseparated roof and clean yard waters
○ Inadequate capacity in   – slurry lagoons
                                      – slurry tanks
                                      – effluent tanks
○ Deliberate breaching of lagoon embankments
○ Deliberate discharge-penstocks in effluent tanks
○ Deliberate overflow pipes in effluent tanks
○ Sudden collapse of   – lagoon embankments
                                      – manure store walls
                                      – silage clamp walls
○ Cracks in silage clamp floors
○ Silage clamps with rubble floors sited over land drains
○ Silage stored outside the purpose-built clamp
○ Inadequate supervision of farm workers
○ Machinery re-fuelling using unsupervised gravity feed
○ Overfilling of, and leaks from, unbunded oil tanks
○ Tank valves opened by livestock brushing against them
○ Sheep-dip baths connected to land drains
○ Deliberate emptying of vacuum tankers into ditches
○ Leaking ammonia tanks
○ Overloaded and/or poorly understood treatment plants
○ Vandalism

One of the principal causes of farm effluent handling difficulties is failure to separate out clean waters from the dirty system. Mixed drainage systems are disasters waiting to happen.

THE 'GREEN' RESPONSE

Society has woken up to the importance of the environment, and this will place additional burdens on industry and the consumer alike.

Agriculture will not escape, but it still retains some of the old special status, in that Government financial assistance has been and will continue to be provided toward solving some of the environmental problems (unlike the rest of industry which is expected to meet all its own costs).

Government

The *Royal Commission on Environmental Pollution* [24] made recommendations in 1979, some of which have now been taken up, for example, 'Extended grant-aid to include provision for pollution control'; 'Review the current practice for the storage of chemicals on farms'; 'Grant-aid should be conditional upon water authority consultation'.

However, ten years on, some of their statements look a bit jaundiced when set against current views, for example, 'Reduction of nitrate levels in water would be less costly to achieve through removal by water authorities than by restrictions on agriculture'; 'Pollution from silage making has been a frequent cause of pollution, but it should however be a diminishing problem'; 'Arrangements by which water authorities can obtain advance notification of sheep dipping now appear to be operating satisfactorily'.

The House of Commons Environment Committee in 1987 submitted a report on the Pollution of Rivers and Estuaries in England and Wales, and the Government made some useful responses to this in 1988 [25].

The committee recommended that there should be a far more interventionist and regulatory approach to farm pollution; grant aid should cover waste handling facilities; the *Code of Good Agricultural Practice* should be enforceable; the special Defence provided by s.31(2) (c) of COPA should be repealed.

In response, Clause 103 of the Water Bill currently before Parliament adopts the hitherto inoperative s.31 (4), and the Government is to draft Regulations to cover the construction of new and extended silage and slurry facilities, and intends that the NRA should be able to serve Notice on existing sites to bring the Regulations to bear on them where necessary.

The COPA pollution control powers transferred across to the Water Bill EXCLUDES the special Defence for farmers adhering to 'good agricultural practice'. The Code will not be statutory, but failure to observe it may be considered during legal proceedings.

The 1985 Agricultural Improvement Scheme provided grant aid for effluent storage, but there has been confusion on whether it extended to drainage separation works. 'A.I.S.' was ended in November 1988, but its successor, the Farm and Conservation Grants Scheme (from February 1989) promises to be more related to environmental issues and less to increased production. £50 million have been promised over the next three years.

The Committee recommended that trial protection zones should be set up in selected sensitive catchments under s.31 (5) of COPA.

Clause 104 of the Water Bill provides for the establishment of protection zones, within which specified activities may be forbidden or permitted subject to conditions (the hitherto inoperative s.31 (5) of COPA).

Country Landowners Association

In 1988 the CLA called for a new Government campaign to stop the pollution of rivers by farm waste and silage effluent. The CLA proposed that there should be a series of county-by-county clean-ups, masterminded by ADAS. Every farm within a target county would be visited by ADAS staff who would provide free advice and offer a temporary high rate of grant for remedial work. It was suggested that if a farmer rejected the advice and was subsequently prosecuted for pollution, his refusal could be brought to the attention of the Court.

Unfortunately, the CLA plan was not taken up.

National Farmers Union

In September 1988 the NFU circulated their members with a fact sheet drawing attention to the rise in farm pollution incidents, pointing out that farmers who polluted rivers were increasingly likely to face prosecution, and that the Government were drafting new regulations on silage clamps and slurry stores.

The NFU urged its members to take action immediately to prevent further pollution.

To assist its members in getting their effluent systems sorted out the NFU has secured a number of special discounts for members on a range of equipment to help solve some common pollution problems.

Water Authorities

Publicity. In recent years water authorities have showered advisory leaflets on the farming community; they have issued press releases; they have taken part in local television and radio programmes; they have organized displays at agricultural shows and have given talks to farming groups. Special campaigns have been launched.

South West Water produced a Video, 'Pollution—Together we can beat it', highlighting the main causes of pollution and its effects, together with the efforts being put in by South West Water and the farming community.

Special surveys. Concentrated surveys of farms have been carried out by South West Water and Yorkshire Water.

South West Water found that on average, first visits revealed 25% of farms to be polluting, 25% a pollution risk, and 50% satisfactory.

Yorkshire Water found 22% polluting, 24% a pollution risk, and 54% satisfactory.

In Yorkshire the majority of polluting farms were found to be tenanted rather than owner-occupied, and the recurring theme was that, in the tenant's view, the cost of adequate effluent disposal systems should be borne by the landlord, and the opposite view was expressed by the landlords, resulting in stalemate.

YWA campaign. Yorkshire Water's Farm Pollution Campaign of 1988 [32] has been tentatively judged as moderately successful in that the summer of 1988 saw a slight decrease in farm pollution incidents as compared with the summer of 1987, even though the total number of incidents from other sources had risen (again).

Incidents other than agricultural were 15% up, agricultural incidents were 1% down. More significantly, farm incidents classified as severe, which were 21% in 1987, were down to 14% in 1988.

Severn–Trent campaign. Severn–Trent Water have supplemented the 'stick' with a 'carrot' in the form of an annual competition with a cash prize for the farmer who has shown the greatest initiative in works to prevent pollution.

Prosecutions. In 1986 there were 128 farm pollution prosecutions in England and Wales, and the average fine per offence was £276.

In April 1987 the Lord Chancellor was reported in *The Magistrate* as saying, 'I am justified in drawing the attention of all courts to the need to impose in this category of offence fines high enough to establish that this type of illegal act can be seen not to pay... may I suggest, therefore, that for this category of offence you should regard the maximum fine £2000 as your starting point, reducing that amount according to whatever mitigating factors exist'.

In 1987 there were 225 prosecutions, and the average fine fell to £205.

At the time of writing, national figures were not available for 1988, but in Yorkshire the average fine had at last begun to rise (£630) with a maximum of £1500 (for a fifth offence). In one case a fine of £1000 was secured for a first offence, but this was subsequently reduced on appeal to £500.

CONCLUSIONS

The final warning
In the Foreword to the WAA/ADAS report for 1987 [8] John Selwyn Gummer, then Minister of State for Agriculture, said that the 1987 increase in water pollution incidents caused by farmers was not acceptable and could not be dismissed as merely the result of difficult weather conditions. He concluded that the polluting farmer damages his neighbours as well as the watercourse. They suffer from his failures, and he undoes all the good which their proper management of farm waste has achieved. '1988 must be the year when the farming community gets really tough with pollution.'

At the 1988 Staffordshire silage conference Mr Gummer said, 'We either make sure that we solve these problems ourselves, voluntarily within the code and within such restrictions as we have now... or we are bound to find that the rest of the community demands that we be restricted and policed in a way which we would find burdensome and restrictive'.

Stick or carrot?
Severn–Trent's competition and cash prizes did not prevent their 1987 farm incidents figure of 521 rising to 619 in 1988.

At the FBA Winter Conference a farmer complained about yet more beating about the head by water authorities, and yet, traditionally, pollution control officers have tried to adopt a co-operative approach, and have been reluctant to recommend prosecution (witness 225 prosecutions in 1987 against 3890 incidents!).

In 1988 the Board of Yorkshire Water decided to sharpen up their prosecution policy.

The future
Looking at the sorry picture which has developed, it now seems inevitable that farmers will be faced with more legal action, more regulations, and more restrictions.

It is a pity that the actions of the few has produced this scenario for the many.

REFERENCES

[1] Pontin, R. A. & Baxter, S. H. (1968) Wastes from pig production units. *J. Proc. Inst. Wat. Poll. Control*, **67**, 6, 632–643.

[2] Ripley, C. T. (1970) Current trends in farm waste disposal. *J. Proc. Inst. Wat. Poll. Control*, **69**, 2, 174–179.

[3] Wheatland, A. B. & Borne, B. J. (1970) Treatment, use and disposal of wastes from

modern agriculture. *J. Proc. Inst. Wat. Poll. Control*, **69**, 2, 195–208.

[4] Willets, S. L. & Weller, J. B. (1977) *Farm wastes management*, Crosby Lockwood Staples, St Albans.

[5] Mann, H. T. (1975) Aerobic treatment of farm wastes. *J. Proc. Inst. Wat. Poll. Control*, **74**, 5, 560–570.

[6] BAA (1988) Fires involving pesticides—The Rhine incident. *Proceedings of Seminar* Crest Hotel, York 1987. British Agrochemicals Assoc. Ltd., Peterborough.

[7] Stansfield, R. (1983) The Diquat incident at Woodkirk, Yorkshire. *J. Proc. Inst. Wat. Engs. & Scientists*, **37**, 4, 364–370.

[8] WAA/ADAS (1987, 1988) *Water pollution from farm waste 1986, Water pollution from farm waste 1987*. Water Authorities Association, 1 Queen Anne's Gate, London.

[9] Yorkshire Ouse River Board (1961) *Eleventh Statutory Annual Report for the Year ended 31st March 1961*, D. C. North, Leeds.

[10] MAFF *Stock figures: Northern Region*. Ministry of Agriculture, Fisheries & Food, Leeds.

[11] Yorkshire Ouse River Board (1964) *Fourteenth Statutory Annual Report for the Year ended 31st March 1964*, D. C. North, Leeds.

[12] Yorkshire Ouse & Hull River Authority (1966) *First Statutory Annual Report for the Year ended 31st March 1966*, D. C. North, Leeds.

[13] Yorkshire Water Authority (1987) *Water Quality Report 1987,* YWA Leeds.

[14] Clyde River Purification Board (1987) *Report for the Year ending 31st December 1987*, The Director, Clyde R.P.B., East Kilbride, Glasgow.

[15] Solway River Purification Board (1987) *Annual Report for 1987,* The Director, Solway R.P.B., Dumfries.

[16] Hobson, P. N. (1984) Anaerobic digestion of agricultural wastes. *J. Proc. Inst. Wat. Poll. Control*, **83**, 4, 507–513.

[17] Sneath, R. W. (1978) The performance of a plunging jet aerator and aerobic treatment of pig slurry. *J. Proc. Inst. Wat. Poll. Control*, **77**, 3, 408–420.

[18] IWPC (1971) Study tour in the Netherlands. *J. Proc. Inst. Wat. Poll. Control*, **70**, 113.

[19] Scheltinga, H. M. J. (1969) Farm wastes. *J. Proc. Inst. Wat. Poll. Control*, **68**, 4, 403–413.

[20] Riley, C. T. (1968) A review of poultry waste disposal possibilities. *J. Proc. Inst. Wat. Poll. Control*, **67**, 6, 627–631.

[21] WAA (1988) *Waterfacts 1988*, Water Authorities Association, 1 Queen Anne's Gate, London.

[22] Bastiman, B. (1976) Factors affecting silage effluent production. *Experimental Husbandry*, **31**, 40–46.

[23] Mason, P. A. (1988) Dealing with silage effluent: technical opportunities. In: *Proceedings Silage Effluent Conference*, Staffs Agric. Coll., Chalcombe Publications, Marlow, Bucks.

[24] Royal Commission on Environmental Pollution (1979) *7th Report. Agriculture and Pollution*. HMSO, London.

[25] House of Commons Environment Committee (1988) Third Special Report. *Pollution of rivers & estuaries—observations by the Government on the 3rd Report of the Committee in Session 1986–87*. HMSO, London.

[26] Mason, P. A. (1988) Construction of slurry and effluent retaining structures. In: *Proceedings Farm Waste & The Pollution Problem FBA Winter Conf., Stoneleigh.* Farm Buildings Association, N. A. C., Stoneleigh, Warwickshire.

[27] Nielson, V. C. (1988) Appraisal of European codes & recommendations being formulated, Farm Waste Management, Animals & Public Health, Application to the Land & Odour Control. In: *Proceedings Farm Waste & The Pollution Problem FBA Winter Conf., Stoneleigh.* Farm Buildings Association, N. A. C., Stoneleigh, Warwickshire.

[28] HSE (1988) Guidance Note CS19 (July 1988) *Storage of approved pesticides: guidance for farmers and other professional users.* Health & Safety Executive, Bootle, Merseyside.

[29] MAFF (1988) Revised Draft Storage Code of Practice. A Draft FEPA Part III Code of Practice on the Sale & Supply including Storage for Sale & Supply of Pesticides approved for Agricultural Use. Ministry of Agriculture, Fisheries & Food, Pesticides Safety Division, Great Westminster House, London.

[30] MAFF (1988) Revised Draft Code of Practice for the Agricultural & Commercial Horticultural Use of Pesticides. Ministry of Agriculture, Fisheries & Food, Pesticides Safety Division, Great Westminster House, London.

[31] WAA/CACFOA/BASIS (1988) Inspection and Approval of Agrochemical Stores by Water & Fire Authorities for BASIS Registration. Water Authorities Association, London. Yorkshire Water, Leeds.

[32] YWA (1988) Meeting of Water Quality Advisory Group. Agenda & Reports, 3rd November. Yorkshire Water, Leeds.

6

Farm waste and nitrate pollution

M. R. Payne [†]

ABSTRACT

Agriculture is an extensive industry utilizing over 80% of the land surface in England and Wales. In recent years a number of far-reaching changes in management practices have been introduced, particularly in animal husbandry. Water undertakers use agricultural land as a catchment area to supply to the ever-increasing urban populations. The density of population in the UK means that intensive farming and water catchment areas largely overlap. This dual use of land brings the practice of agriculture into conflict with those responsible for maintaining the quality of water supplies. The two uses are regulated by different legislation, different Government departments and different national policies.

Water supply for human use may be regarded as the key use to which we put our water. It is only one aspect of environmental water quality. Our society is placing increasing value on maintaining and restoring the quality of its environment, a trend of which agriculture must be mindful.

Agriculture forms a continuum over much of the land's surface, compared with the relatively small and often isolated sites occupied by other industries. Most fundamental is agriculture's integration with the natural environment. Agriculture consists of the exploitation of features of the natural environment to produce food and clothing for human beings. In countries where agriculture is well developed, it has largely become the natural environment of the land surface. Water draining from agricultural catchments will always differ from the catchments where other forms of land use prevail.

The integration of agriculture and the natural environment provides an interface with water resource management almost as extensive as the area of agricultural land. Water draining from agricultural land does so naturally in a largely uncontrolled and uncontrollable manner. Comprehensive management of the water resource as it leaves a particular production in the agricultural context could only be achieved by full control of agricultural practices. The relatively small number of point sources from which discharges from agricultural buildings occur are susceptible to conventional control, but an entirely different concept is required in the context of diffuse pollution arising from agricultural land.

KEY WORDS   Agriculture   Water   Resources   Pollution

INTRODUCTION

Agriculture is an extensive industry using over 80% of the land surface in England and Wales. It has continued in its fundamental form of large numbers of small units for many

[†] Pollution Adviser, Agricultural Resources Department, National Farmers' Union.

centuries, although in recent years a number of far-reaching changes in management practices have been introduced, particularly in animal husbandry. Nonetheless, the industry continues to be organized on a fragmented basis with relatively large numbers of small businesses providing an intimate contact with the land on which production takes place. Water undertakings use agricultural land as a catchment for the commodity they have been established to supply to the ever-increasing urban populations. In the absence of conurbations typical of an industrial society, comprehensive water undertakings would be less in evidence, and local provision for water of adequate quantity and quality could be still be possible. The density of population in the UK means that intensive farming and water catchment areas must largely overlap. This dual use of land as both the catchment from which to derive water supplies and as a production medium is largely responsible for bringing the practice of agriculture into conflict with those responsible for maintaining the quality of water supplies. The two uses are regulated by different legislation, different Government departments, and different national policies, which demonstrates only limited co-ordination.

While water supply for human use may be regarded as the key use to which we put our water, it is only one aspect of environmental water quality. Our society is placing increasing value on maintaining and restoring the quality of its environment, a trend to which agriculture must not be blind.

Agriculture can be identified as forming a continuum over much of the land's surface as compared with the relatively small and often isolated sites occupied by other industries. Over and above this, and more fundamental, is agriculture's integration with the natural environment. A typical industrial enterprise is isolated from the outside environment by walls, floors, and roofs, and operates in a regulated environment. It may use natural raw materials, or discharge its waste, but it functions largely independently of nature. In contrast, a farm depends on soil, rain, sunlight, and atmospheric gases for its productivity. While any or all of these can be artificially supplemented or replaced in certain forms of agricultural practice, agriculture consists fundamentally of the exploitation of these features of the natural environment to produce food and clothing for human beings. In countries where agriculture is well developed, it has largely become the natural environment of the land surface, a fact which is both well recognized and much valued in the visible effects of agriculture as manifested in the landscape of the UK. The effect continues into the less visible effects of the environment, and as a consequence, water draining from agricultural catchments will always differ from catchments where other forms of land use prevail.

The integration of agriculture and the natural environment provides an interface with water resource management almost as extensive as the area of agricultural land itself. The bulk of water draining from agricultural land does so naturally in a largely uncontrolled and uncontrollable manner, in contrast to the typical industrial situation. Comprehensive management of the water resource as it leaves a particular production unit may be a realizable objective in the industrial situation. In the agricultural context, it could be achieved only be full control of agricultural practices. The relatively small number of point sources from which discharges from agricultural buildings occur are susceptible to conventional control, but an entirely different concept is required in the context of diffuse pollution arising from agricultural land.

RESOLVING THE COMPETING CLAIMS OF
WATER RESOURCES IN AGRICULTURE

Where the unrestricted practice of agriculture present difficulties in terms of quality for water resources, a policy must be evolved to resolve the conflict. The options available fall into four categories:

(a) No environmental change tolerated.
(b) No environmental pollution (change with unwanted effects) tolerated.
(c) Limited environmental pollution tolerated.
(d) Virtually unlimited environmental pollution tolerated.

There is an important distinction to be recognized between categories (a) and (b). This is set out in the Tenth *Report of the Royal Commission on Environmental Pollution* (RCEP) [1], which finds the following a useful definition of pollution:

'The introduction by man into the environment of substances or energy liable to cause hazards to human health, harm to living resources and the ecological systems, drainage to structures or amenity, or interference with the legitimate uses of the environment.'

The Commission then goes on to add:

'It is implicit in the above definition that pollution is not simply the presence in the environment of an alien substance or other unnatural disturbance; there must also be an unwanted effect. Substances introduced into the environment become pollutants only when their distribution, concentration, or physical behaviour are such as to have undesirable or deleterious consequences. For many substances whether a particular discharge or emission is considered to be pollution depends not only on the nature of the substance, but also on the circumstances in which it occurs and often on the attitude of people affected and on value judgements.'

In the early days of the Industrial Revolution, little effective control on water pollution existed, and the policy operated therefore fell mainly into category (d). Since that time, the control of pollution has received more attention, and it is no longer acceptable to discharge polluting materials in an uncontrolled manner. However, pollution still exists and is tolerated or 'controlled' via trade effluent consents for industrial or commercial discharges, and for releases from public sewage treatment works. Therefore, the present policy falls within category (c). Category (a) is not an available option in the UK since some change in the environment in general and water quality in particular inevitably accompanies significant agricultural and other human activities. The scope for debate is therefore restricted to categories (b) and (c). While present policy falls into category (c), the questions raised under broad subject of agricultural pollution may range rather wider. The problem of nitrates raises interesting questions in this context. As is indicated later in the paper, it is likely that nitrates have no significantly damaging effects. Assuming for the moment that this was to be the case, a logical policy would not require them to be controlled under either categories (b) or (c). Yet there is pressure for their release to be restricted regardless of the dearth of evidence of harmful effects. Such pressure is not easily justified if pollution control policy is to remain in category (c). It would be a large step indeed from the present overall situation to aim for a target set by a policy in category (b), which would have considerable repercussions for industry and water authorities alike.

However, taking now the opposite assumption in the case of nitrates, namely that significant damage does accrue from the enhanced levels which they now attain in agricultural drainage water, the question which would remain to be answered would be whether the damage experienced exceeded that deemed acceptable by the current UK policy (a category (c) policy). If the damage experienced related to human health, it would be likely to be deemed unacceptable, and steps would be taken to control either the pollution or its effects. When the RCEP considered this matter in its seventh report [2], it attempted to balance the respective costs to the agricultural and water industries of attempting to reduce nitrate concentration of water prior to its leaving agricultural land and reducing the nitrate content of water at the point where it enters public supply. Whilst the calculations made by the Royal Commission may now be out of date, the principle of clearly identifying the cost attributing to any chosen policy before a decision is taken is to be heartily commended. This principle has recently been further emphasized in the case of nitrates with a series [10] of the respective costs of farming changes and water industry action. These studies at least partly support the RCEP view that it is cheaper to take water industry action.

THE AGRICULTURAL CONTEXT

It is important to appreciate that the move to larger farms, increased mechanization, the loss of jobs from the countryside, and increasing yields have been achieved at the behest of successive Governments and under the pressure of diminishing real prices and farm profits over past decades. It is only with the most recent changes of UK agricultural minister that Government exhortations to increase production have been muted. The industry is aware that change is upon it, but has yet to be told what society now requires of it. Farmers are understandably anxious about their future livelihoods, and such investment as is taking place is generally oriented toward cutting production costs for the future. Against this background, any pressure for additional restrictions on agricultural practices are unlikely to receive a warm welcome.

Despite its economic uncertainty, the industry has to its credit the creation and maintenance of the UK countryside. It is also vital to the rural economy of many areas of the country. In the present interregnum while agricultural policy is reshaped, it would be helpful if the industry could be reassured that essential environmental objectives will be encompassed by whatever new policy is adopted, and that farming will not therefore be receiving conflicting messages as to the demands society wishes to make of it, nor placed at a disadvantage with respect to direct competitors in Europe. The industry has responded to society's needs in the past and has the ability to do so again if it is given clear guidance as to the demands made of it, and policy instruments used to direct it are shaped accordingly.

CURRENT PROBLEMS

The most apparent agricultural pollution problems experienced by the water industry are those recorded in the annual statistics collected by the Ministry of Agriculture from the ten regional water authorities in England and Wales. Figures for 1974 to 1987 are set out in Table 6.1. From these annual totals, it appears that the scale of farm pollution has increased from 61 incidents in 1974 to 3890 in 1987, a rise of such proportions that it is hard to accept the figures as an objective assessment. Changes within the agricultural industry cannot explain such a rapid increase, and it appears that other factors are tending to distort the number of incidents reported. When it is remembered that it was in 1974 that the water authorities were first established, and that in the early years they were largely preoccupied with industrial as opposed to agricultural pollution, it is possible to understand that the figures shown might have arisen from a shift in emphasis or effort by the water authorities. Figures for pollution incidents of all types show a very similar trend to those for agriculture [8] and may even be increasing marginally faster.

Table 6.1. Total number of reported farm pollution incidents in England and Wales 1974–87

1974	1975	1976	1977	1978	1979	1980	1981	1982	1983	1984	1985	1986	1987
61	271	273	952	313	1484	1671	2367	2428	2795	2828	3510	3427	3890

The data on numbers of incidents cannot therefore be used to justify a claim that agricultural pollution is increasing relative to other sources. However, even in the absence of any reliable data, the figures are probably adequate to indicate certain trends, and it does seem probable that some forms of agricultural pollution may have increased over the period. The breakdown of the figures between different sectors of the industry are of interest in this context (Table 6.2). When it is considered that silage is largely made for the benefit of cattle, and these figures grouped with those attributable to cows, it can be seen that the dairy industry is implicated to a large and increasing extent. Most of the dairy industry is located in the western half of the country. The five western water authorities (North-West, Welsh, Severn–Trent, Wessex and South-West) between them recorded 2870 pollution incidents in 1987—74% of the total and three times the number of incidents occurring in the eastern five authorities' areas (Table 6.3). A considerable amount of expansion has taken place on dairy units within the period covered by Table 6.1, and there are numerous examples of farms where increases in cow numbers have not been accompanied by similar increases in waste storage and disposal facilities. However, the introduction of dairy quotas in April 1984 will have had the effect of limiting the tendency to further expansion, except where some units buy in quota from other farms. Silage itself may pose a special problem. The recent economic pressure on the profitability of many dairy units has tended to push farmers towards feeding greater quantities of home grown forage, and particularly silage, rather than feedstuffs imported onto the farm. There is clearly the possibility that some producers will attempt to handle more silage with their existing facilities, again tending to increase the risk of pollution.

Table 6.2. Breakdown of farm pollution incidents by source 1981–1987

	1981	1987	Change over 1981/%
Silage	600	1003	+ 67
Cows	790	1970	+ 185
Pigs	267	217	− 19
Other	579	700	+ 21
Total	2236	3890	+ 74

Table 6.3. Breakdown of farm pollution incidents by authority—1987

Anglian	238
Northumbrian	79
North-West	618
Severn–Trent	619
Southern	194
South-West	646
Thames	170
Welsh	716
Wessex	271
Yorkshire	339
Total	3890

In contrast to the increasing figures for incidents arising from cows and silage shown in Table 6.2, it can be seen that the incidents from pigs and other causes have remained relatively constant since 1981. This tends to suggest that the relative position of the dairy industry has deteriorated over the past six years.

Solutions to the continuing problem of chronic pollution from agriculture are not easy to come by. The fragmented nature of the industry, together with the scale of the problem, poses great difficulties. The key to the situation lies partly with education of the individual farmer. With a high proportion of young farmers passing through formal agricultural training, the potential for educating new farmers entering the industry is present and can be used to provide a thorough understanding of the facilities required on a given site, the correct methods of management, and the consequences of its failure. Just as the problem of education of a changing 'stock' of farmers can most easily be tackled through the new intake, so the problem of inadequately designed facilities must also be approached. Once the need for properly designed facilities is realized, it is not particularly expensive to design adequate pollution control features into a new installation. However, modification of existing facilities is a costly exercise and would stretch the resources of some farms to breaking point. It is therefore necessary to provide substantial grant aid to bring existing units up to a satisfactory standard. While some in water authority and other circles may be critical of grant aid to the agricultural industry, it must be realized that farmers do not have the ability to raise their prices to fund additional capital expenditure. Increased capital funding must be found from within the existing budget, and it is frequently the case that insufficient funds are available.

A further avenue which might usefully be explored is the provision of better co-ordinated advice to farmers by the Ministry of Agriculture's advisory service and water authorities. At present there is a major division of responsibility between the ADAS officer, whose prime duty is to the farmer and whose organization handles grant-aid matters, and the water authority officer whose chief responsibility is to the water authority and who is charged with controlling any pollution which may arise from farm works. There is a need to bridge the gap between these two groups of officers with individuals able to command the respect and trust of farmers, who possess an understanding and knowledge of agriculture, and with a full knowledge of grant-aid matters. It is pleasing to note that both the Welsh and Wessex Water Authorities have appointed farm liaison officers in an attempt to fill this need, and that some water authorities, notably South-West, have actually organized joint visits to farms by their officers together with members of ADAS. Several other water authorities have mounted campaigns stressing co-operation with the farmer.

The NFU has recognized that the farming industry must accept the need to improve its record on pollution. It therefore launched its own pollution campaign in 1988 to impress this upon farmers, and to create a suitable climate in which farmers could be persuaded to improve their waste handling facilities. The NFU negotiated special discounts for its members with suppliers of pollution equipment which helps to reduce the cost of improvements. The campaign also calls on Government to improve both the scope and rates of grant aid for pollution control work, and for a return to free ADAS advice on detailed solutions to farm waste problems. It is pleasing that the Government has responded to the first call, although it has yet to act on the second. Both these points were supported by the House of Commons Environment Committee in its 1987 report on river pollution [7]. The NFU has emphasized to its members that legislative restraints will increasingly tighten if the industry, along with other polluters, fails to improve its record, and sincerely hopes that a concerted effort by all interested parties will bring worthwhile results.

NITRATES

In recent years, concern has been felt about rising levels of nitrates in both rivers and groundwater. It appears that roughly 75% of this nitrate originates from agricultural land, and that concentrations tend to be higher in the south and east of England than in the north and west. Concern has been expressed particularly in respect of possible human health hazards, namely methaemoglobinaemia (a form of blue baby disease) and stomach cancer. Nitrates in water have been positively implicated in blue baby disease, but recognition of the problem has led to its avoidance in the UK, where there have been no suspected or confirmed cases whatsoever since 1972 in spite of some high levels of nitrate reached after the 1976 drought. Where the disease has occurred (only 13 cases in the UK) it has been in bottle-fed infants and in association with other factors such as low stomach acidity and high bacterial populations. These circumstances have occurred almost exclusively where well water has been used rather than public supplies. So far as the link with stomach cancer is concerned, this is hypothetical. It is relevant to note that stomach cancer rates are falling in all developed countries, and that within the UK, stomach cancer rates are lowest in East Anglia, where nitrate levels in drinking water are highest, and highest in areas where nitrate concentrations in water are low. Recent research results suggest that the probability of a link between stomach cancer and nitrates is small [3].

There are also some indications that problems may be caused for nature conservation in certain situations by nitrate in water, although the Royal Commission on Environmental Pollution [2] considered that this was not a major difficulty within the UK. This view was reinforced by the report of the Nitrate Co-ordination Group [9], and it is now becoming more widely appreciated that most of the relatively few fresh water eutrophication problems in the UK are linked primarily to phosphate levels, as for example in the Norfolk Broads. Nor are marine eutrophication problems identified in UK waters.

Nitrate release from agricultural land is sometimes attributed to direct run-off or leaching of fertilizer nitrogen. The evidence indicates that this is not the case, and work carried out at Letcombe Laboratory [4] using N-15 as a tracer has demonstrated that the actual loss of fertilizer nitrogen within four years of application is quite small. This is supported by other research findings. The main release of nitrates takes place in the autumn and winter, and it is now understood [5] that this consists largely of nitrogen mineralized from the soil's organic content (which can typically contain 7000 kg of nitrogen per hectare). This has important implications for approaches to the problem, and demands a fundamental examination of soil processes under agricultural management.

One of the main effects of agriculture in managing the land for the greater production of food lies in an increased nutrient flux. This derives from both the addition of nutrients by man to replace those removed in products taken from the land, and in the liberation of nutrients from the organic content of the soil through oxidation after tillage operations. As a consequence of its solubility, nitrogen in the nitrate form tends to be leached from the topsoil in water, while that in the ammonia form is lost to the atmosphere as gas. Virtually any form of agriculture is likely to increase the quantities of nitrogen lost from the soil, although it is only in the last decade or so that concentrations of nitrate in both rivers and ground water have started to cause serious concern.

The problem as experienced in the UK appears to be fundamentally an interaction between rainfall pattern and the quantities of mineralized nitrogen present in the soil after the harvest of other crops. In many other countries, the problem derives mainly from disposal practices of animal wastes. Under UK conditions, there is normally a soil moisture deficit in agricultural land during the summer, and as a consequence little rainfall leaches through the soil to be lost as drainage water. However, with the onset of autumn and winter, soil moisture deficits disappear, and excess rainfall is carried into rivers and aquifers. Where excess winter rainfall is high, nitrate tends to be highly diluted, and the concentration in water supplies does not approach limits currently in operation. However, in areas where excess winter rainfall is less, a similar quantity of nitrate may be less diluted and thus achieve higher concentrations. It appears that the total quantity of nitrate carried by rivers is relatively similar when considered in relation to the total agricultural area of the catchment in both the east and west of England.

The other factor governing the amount of nitrate leached is the quantity of nitrogen mineralized to nitrate which is available in the soil in autumn and winter. Under arable farming systems producing crops such as cereals, there is very little crop uptake of nitrate after June. However, mineralization of nitrogen from the soil's organic reserve remains high in the late summer under the influence of soil temperature, and a substantial accumulation may be present when excess rainfall begins to drain from the land.

Measures intended to limit the application of nitrogenous fertilizers to agricultural land have been suggested as a means of tackling the increasing quantities of nitrate being leached. It appears highly unlikely that such measures could be very effective owing to the role played by the reserve of nitrogen contained within the soil's organic matter. This provides a buffer of immense proportions, the depletion of which could not be countenanced as it would imply destruction of soil fertility. A valuable experiment has been conducted at Rothamsted Research Station where a field has remained unfertilized for over 100 years. During this time, about 2500 kilograms of nitrogen have been lost through leaching, and this loss continues at a rate of about 15–20 kg per annum.

It must be recognized that agriculture is now operating at a higher intensity of both inputs and outputs than has been the case in the past. This 'high pressure' system inevitably imposes greater strains on a process which is inherently leaky. As knowledge of soil processes grows, it is likely that forms of agricultural practice which tend to retain nitrogen in the soil will be better identified, and increasingly introduced for the benefit of both the agricultural industry and water resources. Such practices may include:

(a) Applying the correct level of nitrogen at the recommended timing.
(b) Sowing winter cereals to provide a growing crop to take up residual nitrate in the soil profile in the autumn and thus reduce leaching losses.
(c) Sowing winter cereals as early as possible to encourage yet more of the residual nitrate in the soil to over-winter in the crop where it will not be vulnerable to leaching loss.
(d) Avoiding unnecessary autumn seed bed nitrogens. Autumn nitrogen applications are agronomically desirable in very few if any situations, and there is certainly a case for further reductions to take place.

All these steps relate to positive soil and crop management to reduce the availability of nitrogen for leaching. Such steps have the potential to make a significant contribution to reducing nitrate losses from agricultural land, and they are likely to be more acceptable and perhaps more effective than crude attempts to limit nitrogen inputs.

Over recent months Government decisions on methods of tackling the nitrate problem have begun to emerge. It is now clear that the Government favours a mixed solution, adopting agricultural or water industry methods as appropriate in individual catchments. The criteria to be used in selecting solutions have not yet been announced. It is Government policy to introduce water protection zones whereby farmers are asked to restrict their activities in return for appropriate compensation—probably in recognition of the role that national policy has had in creating the problem. Who will pay the compensation is not yet clear. Water protection zones are to be introduced on a voluntary basis initially, which will greatly reduce policing problems, although compulsory powers will be available in reserve. The type of agricultural restrictions which might be considered seem to me likely to include:

(i) Avoidance of bare land over winter.
(ii) Switches away from problem crops such as oilseed rape and potatoes.
(iii) No ploughing of grassland.
(iv) Conversion of arable land to reasonably low intensity permanent grass.

It must be remembered that much of the current popular interest in nitrates stems from the implementation in the UK of European legislation regulating the nitrate content of drinking water. The maximum limit introduced, 50 mg/l, is half that previously operated in the UK. Attempts by the House of Commons Select Committee on the Environment to identify the basis on which the EC level was determined have revealed an outdated scientific basis for it [7]. Medical advice to British Ministers has been that an average level of 80 mg/l is safe, and in this context the farming industry finds it hard to understand why the UK is adopting a standard of 50 mg/l. Until a rational case can be demonstrated, farmers may tend to feel victimized as a result of pressure by ill-informed protagonists of draconian measures. In the meantime, it is recognized that a legal precedent has been established. Until the law is changed, it must be observed, but farmers must not be penalized for the unforeseen side effects of past Government decisions to expand agricultural production.

CONCLUSIONS

As one of the country's largest industries, it is not surprising that some pollution should stem from agriculture, although increases in the number of pollution incidents do not justify some of the concern expressed. There has been little progress in combating acute pollution problems in recent years, and greater efforts will be needed in the future. The NFU is ready and willing to play its part in this, and urges farmers to take up the increased rates of grant aid available for improvements to their facilities. The example of nitrates has been considered in the realm of chronic pollution, and the close relationship of the industry with the natural environment in the form of soil processes identified. This makes it particularly tricky to tackle some agricultural problems.

In tackling these agricultural problems, two essential points stand out, firstly that agriculture is an industry which is indirectly managed by the state, and that policies need to be designed for the direction of the industry which better integrate society's diverse demands than has been the case in the past. Secondly, there are nearly 250 000 agricultural holdings in England and Wales alone, representing a very wide diversity of situations from agri-business farms covering thousands of hectares to subsistence small-holdings. Such an industry cannot be effectively regulated in the absence of a reasoned case that can achieve a degree of consensus between the regulators and regulated. The nature of the industry requires incentives and persuasion in preference to, but not to the exclusion of, prosecution and criticism.

REFERENCES

[1] Royal Commission on Environmental Pollution (1984) Tenth Report. *Tackling pollution—experience and prospects.* HMSO, London.
[2] Royal Commission on Environmental Pollution (1979) Seventh Report. *Agriculture and pollution.* HMSO, London.
[3] Forman, D., Al-Dabbagh, S. & Dole, R. (1985) Nitrates and gastric cancer in Great Britain. *Nature,* **313**, 620–625.
[4] Dodwell, R. J. & Webster, C. P. (1980) *Lysimeter studies of the fate of fertilizer nitrogen in a shallow arable soil overlying chalk.* ARC Letcombe Laboratory Report, 50–51.
[5] Barraclough, D., Geers, E. L. & Maggs, J. M. (1984) Fate of fertilizer nitrogen applied to grassland, 2. Nitrogen-15 leaching results. *Journal of Soil Science,* **35**, 191–199.
[6] Water Authorities Association/ADAS (1987) *Water pollution from farm waste.* Water Authorities Association, 1 Queen Anne's Gate, London.
[7] House of Commons Environment Committee (1987) Third Special Report. *Pollution of rivers and estuaries.* HMSO, London.

[8] Beck, L. This Symposium.

[9] Department of the Environment (1986) *Nitrate in water*. Report of the Nitrate Co-ordination Group. Pollution Paper No 26, HMSO, London.

[10] Department of the Environment (1988) *The nitrate issue*. HMSO, London.

7

The impact of intensive dairy farming activities on river quality: the Eastern Cleddau catchment study

K. Schofield, BSc, PhD [†], **J. Seager, BSc, MIBiol, CBiol** [†], and **R. P. Merriman, BSc, MIBiol** [‡]

ABSTRACT

Pollution from farm wastes has been one of the principal causes of deterioration in river quality in recent years. However, little is understood of how, and which, farming activities affect the chemical and biological quality of rivers. To address this problem, a field study has been initiated in the Eastern Cleddau catchment, West Wales, to investigate relationship between land use, farm waste management practices, and river quality.

Water quality of small tributaries in this area is poor, and intensive chemical monitoring has shown that discharges from farmyards are a major source of pollution. Rainfall has been shown to exacerbate this effect either through field run-off or wastes washing directly from the farmyards. The status of benthic macroinvertebrate communities in these tributaries is poor and, typically, only a few pollution tolerant species are present.

Small tributaries from farms appear to affect both the water chemistry and biology of larger water-courses, although this effect may be quite localized.

Future research will look at ways to reduce the impact of farm wastes on river quality. This will provide the basis for producing recommendations and guidelines for catchment management aimed at minimizing pollution of rivers from farming practices.

KEY WORDS   Farm pollution   Eastern Cleddau   River quality   Land use   Macroinvertebrate community   Ecotoxicology

INTRODUCTION

Recent years have seen a significant deterioration in the quality of many rivers in the United Kingdom which were previously of good or excellent quality [1]. One of the causes of this decline has been the annual increase in the number of farm pollution

[†] Water Research Centre (1989) plc, Medmenham Laboratory, Henley Road, PO Box 16, Marlow, Bucks, SL7 2HD.
[‡] National Rivers Authority, Welsh Region, Llys Afon, Hawthorn Rise, Haverfordwest, Dyfed, FA61 2BQ.

incidents. Figures recently published by the Water Authorities Association and Ministry of Agriculture, Fisheries and Food [2] show that in the nine years from 1979 to 1988 the number of incidents increased by 179%. Perhaps more alarming is the fact that despite considerable efforts by the water authorities to inform farmers of the potential problems caused by the management and disposal of farm wastes, the number of reported pollution incidents increased by 6% during the last year alone.

Although trends vary from one area to another, the problem is most severe in the wetter, western regions of the country where intensification of dairy farming has resulted in increased waste management problems. To understand the relationship between river quality and farm management practices, the Water Research Centre in collaboration with the water undertakings has established a case study in South West Wales. The objectives of this study are to identify those farming practices which have a significant adverse effect on river quality, quantify their impact, and recommend guidelines and operating procedures aimed at minimizing pollution of receiving watercourses.

STUDY AREA

The area of investigation is the Eastern Cleddau catchment in Dyfed, South West Wales (Fig. 7.1). The predominant bedrock of this area is Ordovician consisting of mudstones, shales, sandstones, and grits which are overlain by till and other glacial deposits. Main soil types of the area are of the Powys Barton series which are deep and generally permeable [3]. Rainfall in the catchment is high (1422 mm in 1988), making the land particularly suitable to dairy farming. The combination of intensive farming practices, high rainfall, and low soil moisture deficit, produces conditions where the risk of land run-off and discharge of farm waste is high. This has resulted in a considerable number of river pollution incidents in the area over recent years.

Other than farming activities, there are no forms of industry within the catchment. Few villages exist and sewage disposal by the resident farmers is via individual septic tanks. The only other potential source of pollution to the river system is from fish farming, although discharges from these establishments are subject to control through water authority consents.

FIELD STUDIES

Catchment wide surveys 1988

Chemical quality of the rivers of the Eastern Cleddau catchment has been evaluated through regular sampling at 23 sites. Measurement of biochemical oxygen demand, dissolved oxygen, and ammoniacal nitrogen suggested that most of these rivers were chemically of good quality at the time of sampling. The only site found to be significantly polluted was Clarbeston Stream (Fig. 7.2), and this tributary was chosen for further intensive study. Biological sampling at 21 sites was carried out during August and September 1988. Benthic macroinvertebrate community structure was assessed by three-minute 'kick' sampling [4]. Biological Monitoring Working Party (BMWP) [5] biotic scores were calculated as an indicator of biological quality at each site. Scores are ascribed to invertebrate families according to their relative tolerance to organic pollution. The lower the score the more likely a site is to be organically polluted.

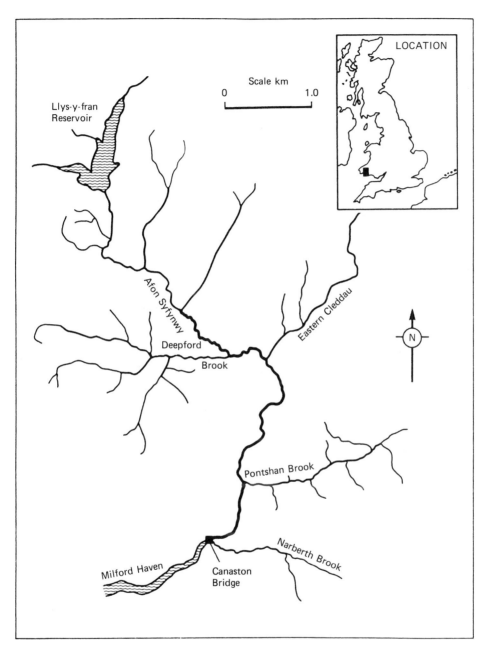

Fig. 7.1. Major rivers of the Eastern Cleddau catchment.

Risk assessment of slurry spreading was carried out by the Soil Survey and Land Research Centre. Mapping within the Deepford Brook sub-catchment enabled a soil series list to be produced. Main soil types were then used in conjunction with climatic and topographic features to classify the suitability of the land for slurry spreading. The work was extended to cover the Eastern Cleddau through interpretation of aerial photographs.

Deepford Brook sub-catchment

An intensive study has been carried out within the Deepford Brook and the Clarbeston Stream (Fig. 7.2). The headwater of Clarbeston Stream receives discharges from drainage pipes which carry parlour and yard washings to the stream channel from four farms. The stream then flows south, passing another farm before reaching the Deepford Brook.

To investigate the relationship between farm management practices and water quality within the catchment, the daily practices on the five farms have been monitored. Co-operation with the farmers has enabled detailed field maps to be produced. Daily diaries are kept as a record of farming practices around the farmyards and on the fields. An inventory of all veterinary products and chemicals used on farm premises has been compiled.

Farmyard slurry has been collected and analysed at various times of the year to evaluate changes in its nutrient content and pollution potential.

Rainfall has been monitored with a 0.5 mm tipping bucket rain gauge. Water depth at two of the three monitoring sites was logged at 15 m intervals and used to calculate discharge.

Water quality was monitored by both spot chemical sampling and with the use of continuous monitoring equipment. Temperature (°C), pH, dissolved oxygen (% saturation), and total ammoniacal nitrogen (NH_4) were measured at 15 minute intervals.

Macroinvertebrate community status was assessed by quantitative sampling at the three monitoring sites. A Surber sampler [4] was used at each site to take 5 random samples from riffle areas. A known area of stream bed (0.1 m^2) was disturbed to a depth of 5 cm and all animals were preserved. After sorting, the animals were identified (to species level in most cases) and densities calculated. Non-parametric statistical analysis was carried out on selected taxonomic groups, using the Mann-Whitney U test [6] to compare invertebrate densities in samples taken from both the upstream and downstream Deepford Brook sites.

RESULTS

Catchment-wide surveys

Pollution risk assessment maps for slurry spreading based on soil type, slope, and proximity to watercourses within the catchment suggest that there are significant areas within the high risk category (Fig. 7.3). These areas are particularly unsuited to slurry application throughout the wet winter months when pollution risk is greatest. Table 7.1 shows the results of the chemical analysis of cattle slurry. The high BOD and ammoniacal nitrogen levels recorded clearly demonstrate the pollution potential of slurry incorrectly applied to land. The results of the catchment-wide chemical and biological surveys [7] suggest that the quality of the major rivers within the Eastern Cleddau catchment was good at the time of sampling. Tributaries receiving drainage water from farmyards were found to be of poor quality.

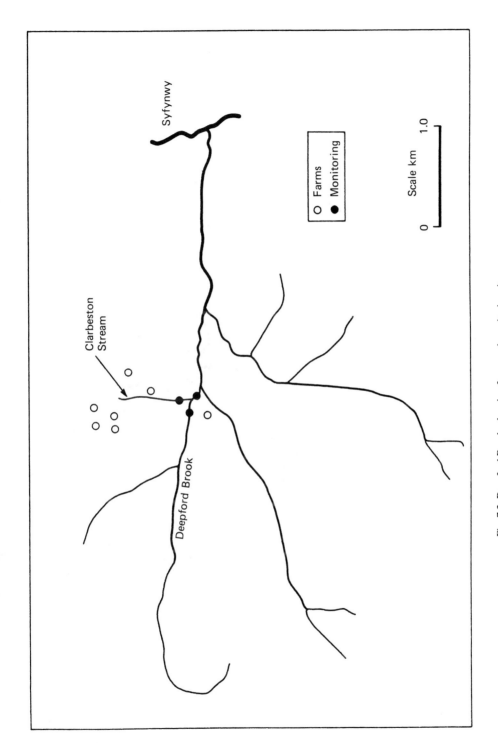

Fig. 7.2. Deepford Brook, showing farms and monitoring sites.

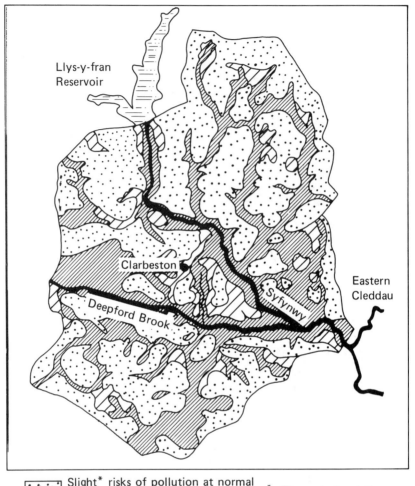

Fig. 7.3. Slurry acceptance map—Eastern Cleddau catchment.

Legend:

Slight* risks of pollution at normal application rate (5 mm/week)

Moderate risks of pollution

Severe risks of pollution

* The scale is relative: Risks are moderate to severe throughout the area due to high rainfall and low moisture deficits.

Deepford Brook sub-catchment

Farming practices

Milking of the dairy herd takes place twice daily, commencing between 6.00–6.30 a.m. and 5.00–5.30 p.m. In the morning this is followed by washing down of the parlour and yard areas. In the evening the parlour alone is washed.

Yearly farming practice routines follow a general pattern: the cows are housed indoors from October to April, during which time slurry is spread on the silage fields. From early

April cows are out during the day only, and by early June the cows are out permanently. During Spring, fertilizer is spread on the fields, and from May onwards silage is cut. The number of cuts varies between two and three, and in intervening periods fertilizer or slurry is spread. Herbicides are used occasionally to control docks.

Table 7.1. Farmyard slurry analysis

Date	Total BOD_5	Ammoniacal nitrogen (mg/l N)	Sulphide (mg/l S)	Total solids (105° C) (%)	Total potassium on solids (%)	Total phosphorus on solids (%)	Total nitrogen on solids (%)
16.3.88							
Farm 1				9.0	2.6	0.53	1.92
Farm 2				8.6	3.0	0.67	2.32
Farm 3				9.4	3.6	0.59	2.16
05.5.88	3350	686					
Farm 4							
19.5.88	12 400	850	15	6.2		0.70	2.3
Farm 2							
30.5.89	5600	1320		9.1		0.77	2.16
Farm 3							
07.7.88	13 600			7.9	2.6	0.51	1.98
Farm 4							

Rainfall

Monthly rainfall summaries from January 1988 onwards are shown in Table 7.2.

Table 7.2. Monthly rainfall summaries for Clarbeston area—January 1988–June 1989

Month (1988)	Rainfall (mm)	Month (1989)	Rainfall (mm)
January	291.5	January	90.0
February	95.3	February	104.2
March	150.9	March	147.3
April	52.1	April	82.2
May	86.9	May	9.2
June	48.0	June	48.5
July	174.4		
August	157.6		
September	88.9		
October	157.5		
November	69.5		
December	50.0		

River chemistry

Data from the continuous chemical monitors show that Clarbeston Stream is grossly polluted with background levels of ammoniacal nitrogen between 3 and 5 mg N/l and peaks as high as 20 mg N/l. Figs 7.4 (a), (b), and (c) show ammoniacal nitrogen levels at the three monitoring sites over a sampling period of 14 days. These data show two major trends: firstly, that Clarbeston Stream affects water quality in the Deepford Brook, and, secondly, that above the normal daily diurnal ammoniacal nitrogen pattern within the Clarbeston tributary are major peaks. External factors therefore appear to be affecting ammoniacal nitrogen levels within Clarbeston Stream.

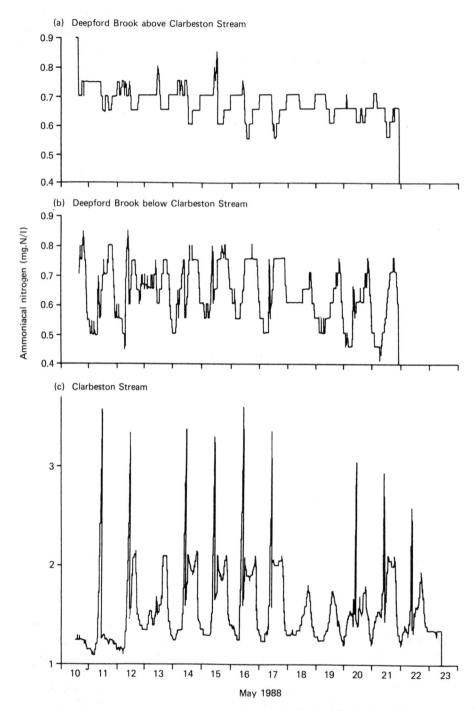

Fig. 7.4. Continuous chemical monitoring in the Deepford Brook sub-catchment.

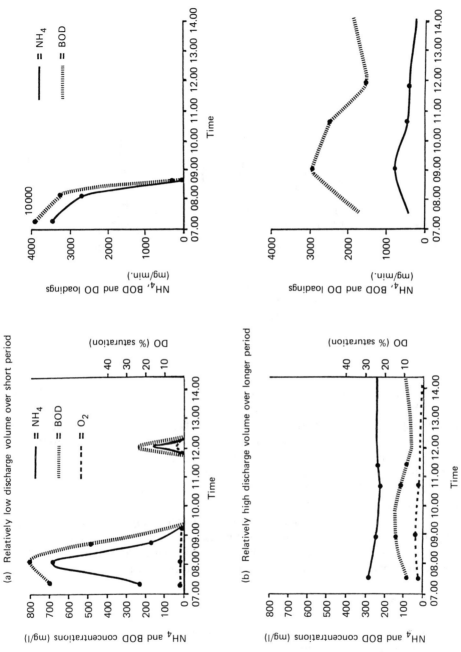

Fig. 7.5. Intensive morning survey—18 April, 1989.

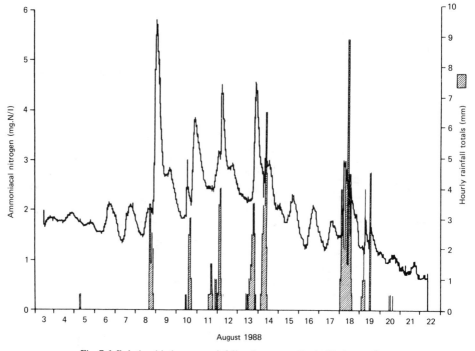

Fig. 7.6. Relationship between rainfall and water quality in Clarbeston Stream.

Intensive spot chemical monitoring of the drainage ditches at the head of Clarbeston Stream has shown that individual farms discharge highly polluted water at certain times of day. Figs 7.5 (a) and (b) show examples from two farms. Discharges were discrete events with BOD levels rising approximately 5-fold and ammoniacal nitrogen levels rising 10–60-fold in a period of 3 h. Levels then decreased again over a further period of 3 h. Associated with the extremely high ammoniacal nitrogen and BOD levels are depressed levels of dissolved oxygen. The timing of these discharges implies that yard and parlour washings are being allowed to enter the watercourses directly.

The effect of rainfall on water chemistry is shown in Fig. 7.6. Rainfall events are generally followed by ammonia peaks 3–4 hours later in the river. The magnitude of the peak appears to be affected by the duration of the antecedent dry period, a rainfall event after a dry period causing a larger peak than one following a period of wet weather.

River biology

The status of the macroinvertebrate community in Clarbeston Stream was found to be poor. Few species were present, although pollution tolerant taxa such as the leech *Helobdella stagnalis* and chironomids were present at high density.

Comparison of certain species populations above and below the Clarbeston input into the Deepford Brook showed that certain groups were significantly more abundant downstream at certain times of the year. These include *Simulium spp*, the chironomids *Brillia modesta*, *Tanytarsini* and 'other Orthocladiinae', and the leech *Helobdella stagnalis*

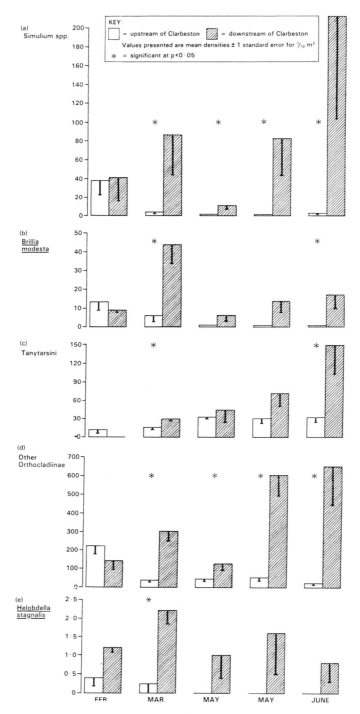

Fig. 7.7. Taxa with increased benthic densities below Clarbeston Stream.

(Figs 7.7 (a)–(e) respectively). Since the physiographic features of these two sites are similar, it seems likely that the differences in invertebrate fauna observed at these sites canbe attributed to the deterioration in water quality caused by the Clarbeston Stream.

Groups known to be more sensitive to organic pollution such as stoneflies (Leuctridae), the Elminthidae beetle larvae (*Limnius volckmari* and *Elmis aenea*), and the gastropod *Potamopergus jenkinsi* were significantly less abundant downstream of the input at certain times of the year (Figs 7.8 (a), (b), (c), and (d)). Other species were apparently unaffected by the Clarbeston input, namely the mayflies *Baetis rhodani* and *Habrophlebia fusca*, the freshwater shrimp *Gammarus pulex*, and certain caddis larvae.

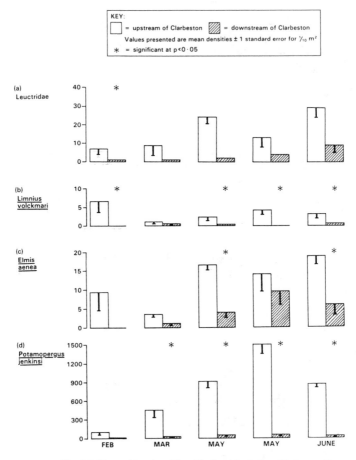

Fig. 7.8. Taxa with reduced benthic densities below Clarbeston Stream.

DISCUSSION

Although it has often been suggested that the deterioration in quality of some of the watercourses in the Eastern Cleddau catchment has been caused by farm pollution, there has previously been little evidence to support this view. Studies undertaken by Welsh

Water have demonstrated that agricultural pollution has detrimental effects on the invertebrate fauna and fish populations in some headwater streams in West Wales. Generally, major effects are not found in larger watercourses [8, 9]. The present intensive study has therefore provided a valuable insight into the effects of specific farming practices on river quality.

Catchment-wise monitoring has shown that most of the land within this area is unsuitable for slurry application at any time other than the dry summer months. Spreading carried out on unsuitable days could potentially lead to pollution of adjacent watercourses owing to land run-off. Although the risk of pollution is high in the catchment, most of the watercourses were found to be of good quality with associated diverse biological communities, thus agreeing with the results of the Welsh Water survey [10]. Certain small drainage ditches, however, were found to be grossly polluted, suggesting that farm pollution could be localized around watercourses draining directly from farmyards and areas of intensive land use. This has been confirmed by the intensive study within the Deepford Brook sub-catchment.

Continuous chemical monitoring of Clarbeston Stream indicated that water quality was poor, with ammoniacal nitrogen reaching levels of up to 20 mg N/l. In addition to these high background levels, a peak of ammonia was seen every morning during a two-week period at around 8.00 a.m. When this observation is compared with the results of the intensive study of the drainage ditches and recorded farming activities, it is apparent that the dominant activity at that time is the washing of the parlour, dairy, and yard areas.

Chemical quality of Clarbeston Stream is strongly influenced by rainfall. Water quality deteriorates after rainfall events and is probably due both to wastes washing from the farmyards and to applied slurry washing from the land. These observations suggest that activities on and around the farmyards are at least partly to blame for the poor water quality in Clarbeston Stream. Chemical monitoring of individual discharges shows that the major point source of pollution is from reception pits designed to contain daily parlour washings. These discharges are highly polluting, and this emphasizes the importance of the water authorities' various campaigns to educate the farming community on correct waste handling procedures and the necessity of separating clean from dirty water before it reaches the containment facilities.

The poor chemical quality of Clarbeston Stream is reflected in the poor biological status of the stream with high densities of only a few pollutant-tolerant species. The observed ammoniacal nitrogen levels of 600 mg/l and dissolved oxygen of 1.2% saturation within the effluent would be directly toxic to all invertebrate species at the point of impact. As these discharges from the farms are diluted, conditions favour taxa which are tolerant of organic pollution such as chironomids. Interestingly, data from the continuous chemical monitor within the lower reaches of Clarbeston Stream record water quality levels which should theoretically support a more diverse fauna. Ecotoxicological work on dissolved oxygen and ammoniacal nitrogen suggests that these levels would not be directly toxic to certain invertebrate and fish species within Clarbeston [11, 12, 13, 14].

However, recent work suggests that low dissolved oxygen levels above the lethal threshold can cause avoidance behaviour in various fish and invertebrate species. Pollution from farming practices could therefore be having a more indirect effect, possibly through avoidance mechanism or owing to sedimentation effects from the high suspended solid loadings. Suspended solid levels of 1 160 mg/l have been measured in discharges from the farms. As these settle out within Clarbeston Stream, they could have an effect through physical disturbance or sedimentation. This may affect habitat heterogeneity, distribution of invertebrates [15, 16, 17, 18], and spawning areas for fish [19].

It would seem therefore that results to date indicate a link between daily farmyard practices and poor chemical and biological quality in the Clarbeston Stream. Evidence from this study suggests that depressions in the water quality of Clarbeston Stream are also reflected in the Deepford Brook. On each occasion that an increase in ammoniacal nitrogen was detected in Clarbeston it could also be detected in the Deepford Brook below the input. The effect is probably quite localized as dilution of the input is considerable. The impact was also observed biologically: pollution-tolerant species were more numerous below the input, whilst more sensitive species were less abundant below the confluence. Generally, the macroinvertebrate benthic community of the Deepford Brook was depressed both in population density and in terms of number of taxa below the Clarbeston input. This suggests that daily inputs of pollutants from the farms via Clarbeston Stream are having a discernible impact on the chemical and biological quality of the Deepford Brook even though the input is relatively minor in terms of volume.

Future research will continue looking into the effects of farm pollution on invertebrate biology. Initially, *in situ* toxicity tests will be deployed to determine mortality rates of selected species, thus providing information on lethal effects of farm wastes. This will be supported by laboratory ecotoxicological work on pulses of specific pollutants such as unionized ammoniacal nitrogen and low dissolved oxygen.

Field deployment of the *Gammarus* 'scope for growth' technique [20] is planned for 1989. This method provides a measure of sub-lethal physiological stress, and has been found to be a sensitive indicator of pollution.

To produce recommendations on ways to ameliorate the impact of farm pollution, research has been initiated to look at cost-effective methods of wastewater treatment. These include the possible use of river corridor buffer zones to prevent land run-off of slurry and the use of reed beds to nitrify the high ammoniacal nitrogen levels associated with yard discharges [21].

CONCLUSIONS

(1)   Water quality in the major rivers within the Eastern Cleddau catchment was generally found to be good, although high ammoniacal nitrogen and BOD levels were recorded in certain tributaries draining land used for dairy farming.

(2)   Intensive monitoring in one such tributary, the Clarbeston Stream, has indicated that certain farming activities such as washings of yards and dairy parlours cause a measurable deterioration in the quality of receiving waters.

(3)   Rainfall exerts a direct effect on stream quality. Significant rainfall events result in peaks in ammonia levels and concomitant depressions in dissolved oxygen, thought to be caused by farm wastes washing off impermeable farmyard surfaces and from land which has received slurry application.

(4)   Changes in the physicochemical properties of watercourses caused by farm pollution appear to be reflected in the status of benthic macroinvertebrate communities.

Tributaries receiving farm wastes were found to be characterized by dense populations of a few pollution-tolerant taxa.

(5) Results from this study will provide the scientific basis for improved catchment management practice. For example, pollution risk maps have been drawn up for the study catchment based on measurements of soil type, soil slope, and proximity of land to watercourses. They will provide guidance to the farmer and pollution control authority for changes in farm management practices to reduce their impact on river quality.

REFERENCES

[1] Department of the Environment (1985) *River Quality Survey, 1985, England and Wales*. HMSO, London.

[2] Water Authorities Association (1989) *Water pollution from farm waste 1988. England and Wales*. Water Authorities Association, 1 Queen Anne's Gate, London.

[3] Rudeforth, C. C., Hartnup, R., Lea, J. W., Thompson, T. R. E. & Wright, P. S. (1984) Soils and their use in Wales. *Bull. Soil Surv.*, Great Britain.

[4] Hellawell, J. M. (1978) *Biological surveillance of rivers*. Water Research Centre, 331 pp.

[5] Biological Monitoring Working Party (1978) *The 1978 National Testing Exercise*. Department of the Environment Technical Memorandum No 19.

[6] Elliott, J. M. (1977) *Some methods for the statistical analysis of samples of benthic invertebrates*. FBA Scientific Publication No 25.

[7] Schofield, K., Whitelaw, K. & Merriman, R. P. (1989) The impact of farm pollution on river quality in the United Kingdom. In: H. Laikari (ed.) *River Basin Management*, Pergamon, Oxford, pp. 219–227.

[8] Wightman, R. (1988) *The results of fisheries and biological surveys to assess the impact of agricultural pollution in five catchments in West Wales*. Welsh Water, Scientific Services, South Western District Internal Report No SW/88/22.

[9] Wightman, R., Jones, F. & Stringer, N. (1989) *The effects of agricultural pollution on headwater fisheries in West Wales, Welsh Regional Division*. South Western District Internal Report No EAW/89/1.

[10] Jones, F. H., Bent, E., Mills, M. & Humphrey, N. (1987) *The biological quality of Welsh rivers (1980)*. Welsh Water, Scientific Services Report No SW/87/13.

[11] Grant, I. F. & Hawkes, H. A. (1982) The effects of diel oxygen fluctuations on the survival of the freshwater shrimp *Gammarus pulex*. *Env. Poll. (Ser A)*, **28**, 53–66.

[12] Williams, K. A., Green, D. W. J. & Pascoe, D. (1986) Studies on the acute toxicity of pollutants to freshwater macroinvertebrates—3. Ammonia. *Arch. Hydrobiol.*, **10**, 1, 61–70.

[13] Arthur, J. W., West, C. W., Allen, K. N. & Hedtke, S. F. (1987) Seasonal toxicity of ammonia to five fish and nine invertebrate species. *Bull. Environ. Contam. Toxicol.*, **38**, 324–331.

[14] Solbé, J. F. de L. G. & Shurben, L. G. (1989) Toxicity of ammonia to early life stages of rainbow trout (*Salmo gairdneri*). *Wat. Res.*, **23** (1), 127–129.

[15] Rosenberg, D. M. & Wiens, A. P. (1978) Effects of sediment addition on

macrobenthic invertebrates in a Northern Canadian River. *Wat. Res.*, **12**, 753–763.

[16] Culp, J. M., Wrona, F. J. & Davies, R. W. (1978) Response of stream benthos and drift to fine sediment deposition versus transport. *Can. J. Zool.*, **64**, 1345–1351.

[17] Wagner, R. H. (1986) Effects of an artificially silted stream bottom on species composition and biomass of Trichoptera in Breitenbach. *Proc. 5th International Symposium of Trichoptera*, Lyons, France.

[18] Reynoldson, T. B. (1987) Interactions between sediment contaminants and benthic organisms. *Hydrobiologia*, **149**, 53–66.

[19] Ringler, N. H. & Hall, J. D. (1988) Vertical distribution of sediment and organic debris in Coho salmon (*Oncorhynchus kisutch*) redds in three small Oregon streams. *Can. J. Fish Aquat. Sci.*, **45**, 742–747.

[20] Maltby, L. & Naylor, C. (1988) *Progress report on the development of biological monitoring procedures for rivers.* Water Research Centre, Report No PRS 1995–M.

[21] Stark, J. H. (1986) *Preliminary results of test tanks to assess the root zone method of sewage treatment.* Water Research Centre, Report No ER 1359–M.

Discussion on Papers 4, 5, 6, and 7

PAPERS 4 AND 5

Mr R. Allcock (Tay River Purification Board), opening the discussion, referred to the problem of the contamination of groundwater used for drinking water by nitrates and pesticides, and said that his own experience had involved work in Yorkshire and Lincolnshire. He asked Mr Headworth what sort of aquifer protection policy his own organization now operated? He said that this was an area of some controversy at present, but when he had worked in Yorkshire there had been an arbitrary system of a 1 km protection area around boreholes. He asked if opinions had hardened during the last few years?

Mr Allcock said that most of the problems in Scotland concerning pesticides and silage were connected with surface waters. Very little of the water supply in Scotland was drawn as groundwater, but there were still problems. One problem was that of sheep-dip being discharged to surface waters.

With reference to pesticides he asked Mr Headworth to comment further on the pesticide levels shown in Table 4.1 of his paper.

With reference to nitrates, Mr Allcock said that the consumption of nitrogenous fertilizer as plant nutrient in the UK was 250 000 t in 1953 and by 1983 had increased to 1 500 000 t. He asked if Mr Headworth would comment on this large increase?

Mr Allcock said that he felt Mr Beck's presentation had been very practical and illustrated the conditions actually found on the farms. He felt more optimistic than Mr Beck had indicated at the end of his paper, in that there was evidence of improvements being made at many farms.

He said that although there had also been an increasing trend of farm pollution incidents in Scotland, he was more optimistic than Mr Beck because there was a high proportion of farms that were not giving rise to any problems at all, and with increased inspections he hoped that would continue.

He asked Mr Beck whether, in England and Wales, any work was being done on peripheral drains for containing silage effluent? Silos were not watertight and they eventually leaked. Peripheral drains were now widely in use in Scotland.

In conclusion, he said that he felt that there might have to be an increase in the number of prosecutions that were taken, particularly for the repetitive offenders. He said that the policy of his own Board was that attempts were made to gain the co-operation of farmers as much as possible, but repeated incidents were resulting in more frequent prosecution. He asked for Mr Beck's views on this aspect.

Mr H. Black (Farmer) said that he had experienced all the problems referred to by Mr Beck, and now they had been overcome. He had spent about £25 000 on a new silage pit, with collection tanks and a system of tank emptying. However, he said that Mr Allcock's comments had been correct—it had a polythene lining underneath it, the drains were designed by an engineer and approved by MAFF, and it had been recently discovered that it leaked and there was a reservoir of effluent underneath the concrete. Fortunately it had been caught before damage was done. A peripheral drain had now been installed.

With regard to pollution and slurry, he said that he had installed a £23 000 system on the advice of three statutory organizations. They had helped design it, and it had achieved a lot of praise and had been monitored with much interest through the construction stage. It consisted of a slurry collection chamber and strainer with a pumping system to lagoon. He said that the three authorities concerned had chosen the fields of the farm that the effluent should be discharged onto. However, before it was ever used, suddenly for no reason that had been explained to him, it had been decided that it might get into someone's water supply, and subsequently a water protection zone had been imposed upon him. He said that the Minister had said in the House of Commons that such zones had never been used, and at this Symposium speakers had referred to them as being a last resort. Through these restrictions he was left with so little ground to actually put out the effluent that it was a continuous problem. Because he had kept thinking that the whole thing was crazy and people would change their minds when they saw his predicament, he had delayed; some had been got out onto the land, but now in March it had overflowed and was overflowing daily, and there was nothing that he could do to resolve it.

Mr Black referred to the EC standard regarding the number of livestock units that should be kept in relation to the area available for the spreading of slurry and effluent. He said that his farm had fallen comfortably within that standard. However, the 'goal posts were moved' because there was now a question of restricting slurry to 20 m^3/acre on vulnerable land. He said that if one had a water protection zone on part of the farm, that automatically meant that one had to increase the amount that had to be spread on the other part of the farm. If incidents of the rules being changed continued, then all the calculations that one had done and all the investment that one had made would come to nought. He concluded that entire catchment areas and entire farms had to be considered and, above all, the authorities had to talk to the farmer—not just impose things upon him. The best acceptable solutions must be jointly determined, even if they were interim solutions in order for time to be given for permanent solutions to be organized in the most acceptable and least environmentally damaging way. He felt that too often officials used the laws as 'a stick to beat people with'.

With regard to his own problem of disposal, Mr Black said that consideration had been given to disposing of his effluent onto his neighbour's land, thinking that he would be grateful to receive it. However, the neighbour had discovered that they also had private water supplies derived from their land and had been terrified to accept it.

Mr R. Aspinwall (Aspinwall & Co.) asked Mr Headworth and Mr Beck whether or not there were problems of pesticide residues in (a) surface waters, and (b) groundwaters?

He questioned Mr Headworth with regard to the status of the sampling from the water authorities who had been kind enough to supply him with results. He asked if the author knew whether the samples were from pumped supplies (public water supplies) or whether they were from the unsaturated zone, because, as he explained, with the nitrate profiles moving through the unsaturated zone, he wondered whether, with some of the more persistent chemicals, there might also in fact be peaks of those in the unsaturated zone?

Mrs E. Warr (Writtle Agricultural College) said that she was concerned that in agricultural education the teaching of 'production' was often divorced from any environmental implications. For example, the production of silage was usually taught separately from farm waste management, and the relationship between the use of chemical additives and the quantity and quality of effluent produced might be ignored.

She asked if the authors had found this to be true, and if so perhaps there was a role for those in the agricultural colleges in trying to link production with the environmental implications.

Mr S. Ross (Clyde River Purification Board) referred to the new conservation and woodlands grant scheme, saying that it seemed to him to be much akin to the 1951 Rivers (Prevention of Pollution) Act in that it referred only to new discharges, with the consequence that little was achieved. He understood that the new grant scheme had been brought in to alleviate the problems from silage and slurry incidents, and he felt that it would not achieve that if it did not take into account the approval of grants for upgrading existing structures.

He said that recently his Board had received enquiries from two farmers looking for grants for repairs to the floors of their units. They had been informed by the Department of Agriculture & Fisheries for Scotland that grants were not available for this.

He said that one thing that he would like to see come out of this Symposium was the Institution making strong representations to the MAFF to extend the grant system.

Mr A. Stevens (Boythorpe Ltd), referring to Mr Beck's paper, said that one thing that should be added to the list of factors which have contributed in the past two years to the increase in farm pollution was that maintenance was required at all times. He said that reference had been made to units that leaked, and he stressed to farmers that quality materials should be used at all times.

He said that even a 50% grant could still mean a farmer spending £20 000–£30 000, and in the present climate that was a lot of money. Farmers were not getting the returns, and this should be borne in mind; perhaps they should be helped more. He said that the 50% grant was actually a lowering of 10% in some instances.

Mr D. A. Burroughes (British Effluent & Water Association) said that in Paper 5 it was recorded that Friends of the Earth drew attention to 300 UK potable supply sources with levels of pesticides in excess of the EC Drinking Water Directive. He asked if the water authorities had checked these reports? Did they refer to surface or ground water, and if so where were the samples taken? Were they likely to be the result of point source or diffuse pollution?

He also asked about the degree to which over-stocking on farms without disposal space for manure and slurry was contributing to farm pollution incidents.

Mr R. M. Walls (West Hampshire Water Company) said that the emphasis so far had been on technical aspects. He was disappointed in not seeing as many farmers present as he would wish, but he was even more disappointed in not seeing many politicians present, because he wondered if this was more a political matter and consequently a matter of money. He felt that although perhaps the technical aspects might need some additional information, by and large one knew technically what was occurring, and really what was needed was money and the political will to do something about it.

Mr A. R. Staniforth (Reading Agricultural Consultants), in a written contribution, said that during the discussion on controlling silage effluent, there was considerable emphasis on the importance of site construction, but little was said about the usefulness of absorbents in reducing effluent. Absorbents could be expected to soak up at least three times their own weight of liquid, and there was good evidence that they improved the fermentation of low dry matter silages. Absorbents did not necessarily reduce the nutritive value of silages when properly used, and they intercepted nutrients which would otherwise go to waste. They obviously had to be matched to the moisture content of the ensiled material and they could not always be expected to eliminate effluent, but their use should be encouraged as an economic method of reducing a serious pollution problem.

PAPERS 6 AND 7

Mr D. Braithwaite (Bishop Burton College of Agriculture), opening the discussion, said that the last two papers had shown clearly that there was a high level of pollution, which was unacceptable, in particular from the dairy industry. He said that this required overcoming, but to improve facilities and provide adequate slurry storage to enable spreading to be carried out at the desired time, it was going to cost a great deal of money. The fact was that since 1975 there had been a continuous decline in real financial income from farming, and predictions were that this situation would get worse before it improved. He therefore wondered who was going to 'pick up the price tag' for such investment? It had already been indicated that the conservation-related grant aid was inadequate.

Mr N. Whitley (States of Jersey Agricultural Department) said that in the past he had taken part in work on the River Torridge, as a member of the ADAS team, and had specialized in dairy farming and grassland management. He said that he wished to endorse the South-West initiative whereby the River Torridge situation had been looked at as a joint venture. He said that it was a problem primarily presented as a result of the intensification of the dairy industry; the district had 20 000 dairy cows, a large proportion of which were fed on silage. It was a high rainfall area and an area of old farms which had 'antiquated' effluent systems. Joint visits had been carried out, and whilst there had been some resistance, in general the problem was accepted in the area. He said that working together had been of definite benefit and hoped that it would develop further in other areas in order to get over the answers to some of the problems which had been presented. He had found the farmers' attitude on the whole had been reasonable—they had accepted that there was a problem, and in many cases the initiative came from the farmers.

As the previous speaker had commented, the funding of the operation was a problem. However, he said that there were many problems that could be solved at low cost. He was pleased that the rate of grant had been increased for such work; any support for such work was going to lead to an improved situation.

Mr Whitley said that the results which had recently been obtained in the River Torridge area confirmed that the actions which had been taken by farmers were being successful.

Mr J. E. Turley (Yorkshire Water Authority) said that he was very interested when a previous speaker had referred to the high nitrate levels in some rivers and water supplies being from sources other than nitrogen-based fertilizers, for example sludge spreading. He said that the use of nitrogen fertilizers could be restricted or even banned, but there was no way that one could stop sludge spreading, both farm sludges and sewage sludges. He said that direct injection into the soil was being increasingly used for sewage sludges and he wondered, if it were used for farm sludges, whether this would reduce the amount of nutrient run-off and possibly avoid unnecessary restrictions?

He said that as an example, in the area in which he worked there was a reservoir which every autumn regularly suffered from an algal bloom. The source of the problem had not been identified, but there was little if any arable farming in the area, so the cause was almost certainly not fertilizers. However, it could well be the result of cow slurry spreading.

Mr M. G. Booth (Wessex Rivers) said that there had been discussion about the benefits of grant aid and the fact that farmers were in a different position from the rest of industry because they did not have control over the price of their produce. He said that in the discussion on Papers 4 and 5, Mr Black had described expenditure that was considerable and had not necessarily produced the answer that everyone was looking for. He said that he did not wish to discuss the rights and wrongs of the particular case, but he knew that Mr Black was a good farmer because he was able to say that he knew that his lagoon was overtopping and he knew that there was leakage of silage liquor. All too often it was found that farms where considerable amounts of money had been spent, with tax-payers' help, to put in systems which should do the job were still causing problems. The fundamental reason for that was that the farmer had put in a system and then neglected it.

He felt that one of the things that needed to be addressed within the farming community was not just a question of where the money came from but of where the culture change came from. It was a culture change to recognize that good farming was not just a matter of yields, but also of looking after wastes and the systems for handling them. The culture change surely could come only from within the farming industry, and one basic area that could promote it was education right at the grass-roots level. He said that several representatives of agricultural colleges were present and that it was not our universal experience that colleges of agriculture welcomed advice or requests to give talks on farm pollution. It concerned him that prevention of farm pollution was not an integral part of the education of young farmers. He hoped that the NFU would put much more pressure to change the culture of the young farmer who was being trained at present.

Mr R. Aspinwall (Aspinwall & Co.) suggested that there was already a technique in farming that went a long way to (a) reducing the high pressure of the system and (b) reducing the nitrate leaching and some of the other problems. That was organic farming. He felt that organic farming had been presented to this conference in an entirely negative way, which had surprised him because there was a great deal of interest in organic farming, not only from so-called fringe people, but also from professional people who were concerned with things like the quality of food and the quality of water.

Mr Aspinwall said that he was no expert, but that organic farming operated on a principle of building fertility in the soil itself—it actually looked after the soil as opposed to just the crop. The building of that fertility was really based on a grass and clover ley which was then used when it was ploughed in as the basic nutrient input to the following crop. Because it was all based on a rotational period, the ploughing-in process usually took place perhaps only every two years, every four years, or maybe even as long as every six years, depending on the rotations which were chosen. If this was done efficiently, much of that nitrogen reservoir and nutrient reservoir was designed to be taken up by the following crop, and the amount of leaching which would occur would be restricted.

He asked Mr Payne why the NFU did not appear to take those points on board and present organic farming as an 'alternative' way of looking at things. He said that surely the conventional way of farming was really the alternative way of looking at things. This rotational process had been well established and, as he understood it, there was good research information to substantiate his remarks.

He said that it was now known that the high pressure agricultural system was causing extreme damage to soil structure through the loss of organic input, and this was an indicator of the stress under which modern agriculture was putting the environment in general.

Mr I. Svoboda (West of Scotland College of Agriculture) said that there had been much mention of the pollution of water from nitrates and farm wastes, but there had been little reference to the problem of malodours. He had been working for more than ten years on the aerobic treatment of animal wastes and had found that such processes had diminished pollution and also had led to an increase in stock numbers because the application rates could be doubled owing to decrease of nitrogen and BOD.

He expressed concern that in his paper Mr Beck had stated that the message to the farmer was 'forget about treatment' and asked if he knew that these types of treatment plants were available?

Mr R. M. Walls (West Hampshire Water Company) said that he wished to make three points about the words that had been used. He referred to the last sentence of Mr Headworth's paper (Paper 4) which talked about 'reducing teaching losses from nitrogen fertilizers', although he thought he knew what it was meant to say. Earlier in the same paper there had been reference to Ministers 'grasping the nettle'—nettles were very good indicators of high nitrogen in soil. Finally, he said that he was sure that the Conservative Party had missed out on an extremely important point—*Phragmites australis* was of course very good at nitrate removal, but its real name was Thatcher's Reed!

Mr S. W. Bailey (ADAS/MAFF) referred to Mr Turlet's comments concerning injection and manures. He said that injection was an option that was often considered, but there were various constraints on its use, particularly on grassland—the same constraints that would apply to the injection of sewage sludge. However, there were situations where it offered various advantages, possibly in reducing run-off and also in reducing odour problems during spreading.

With reference to the Eastern Cleddau study, he said that he was interested to hear that there was included an investigation on the root-zone method. He said that ADAS was also investigating the technique, and two trial reed beds had been established on farms in collaboration with the WRc, Birmingham University and an engineering company. These had been designed to investigate the treatment of dilute waste such as yard run-off. He said that preliminary results suggested that even with those dilute effluents the technique did not achieve much treatment. The ingoing effluent would have to have a BOD diluted down to 1000 mg/l or less before the bed could treat it sufficiently.

He said that this really reflected a larger problem. Yard water was recognized as a problem in causing pollution. Low-rate irrigation was often used as a means of disposal, but there were situations when this was not suitable, for example wet western areas on heavy land. He said that there was a shortage of techniques in some situations for yard water disposal. Reed beds might offer a solution, but he said that he was not over-optimistic.

Mr R. Merriman (Welsh Water Authority) said that Mr Payne had quite correctly stated in his paper that about 75% of all farm pollution incidents were attributable to the dairy sector. This meant that less than 20% of the farmers were actually causing 75% of the incidents. He therefore asked that, given the various deficiencies that had been referred to with regard to the new MAFF grant scheme, should a new scheme be considered, and developed in consultation with the water authorities and the farming unions, tailored specifically to dairy farming? Furthermore, should the water authorities and ADAS be following up work aimed specifically at the dairy industry?

Dr H. Montgomery (Consultant) said that Dr Schofield had referred to reed beds and buffer zones as possible means of dealing with wastes from dairy farms. Specific mention had been made of the nitrification of the amm.N. He said that bearing in mind that each mg of amm.N required about 4.3 mg of oxygen to nitrify it, he questioned whether nitrification in a buffer zone was feasible in the context of concentrations of several hundreds of mg/l of amm.N in the farm effluent.

Mr B. D. Ogden (Yorkshire Water NRA) said that he had listened with interest to the various comments. In his experience, over the last 25 years, of going round farms he had been listening to many of the same problems. Reference had been made to various campaigns and education. He said that many of his colleagues and staff had already done much of this and to some extent one was preaching to the converted. Farmers who were interested would actually do the job. It was the other 90% that the message needed getting over to, and the only way that he could see of achieving that was for individual site visits.

Mr N. C. Oxley (Howard Humphreys) said that he believed that the area where Dr Schofield's work (Paper 7) had been carried out had a very high amount of karstic limestone in it. He wondered to what extent the geology, particularly in a karstic area, was taken into account when slurry wastes were disposed of?

He had taken part in groundwater investigations in the Milton area of Pembrokeshire, where it had been found, on pump testing a number of wells, that there were occasionally high background concentrations of pathogenic bacteria which could be attributed to farm slurry wastes infiltrating to the limestone aquifer.

Obviously in the context of a karstic area he felt that the idea of buffer zones was not really a satisfactory one.

Dr W. O. Jenkins (Construction Industry Research and Information Association (CIRIA)) thought that several important aspects had been raised during the session, in particular the need to discuss things openly between the parties concerned. There was also a need for dissemination of information to the farming community. A third point that had arisen had been the significant reduction in pollution that could occur by the adequate containment of liquid wastes. On the latter aspect he said that CIRIA had developed a research proposal that had been circulated to the WRc, WAA, NFU, ADAS, and the CLA. This proposed the development of a document providing appropriate guidance to the farming community on the construction of such facilities. He felt that this was an opportunity for related farming, regulatory, and manufacturing organizations to come together to help in the dissemination of guidance which would benefit both the farmer and the environment.

Mr C. J. A. Binnie (W. S. Atkins & Partners), referring to Mr Payne's paper, (Paper 6) said that the author had referred to the nitrate standard of 50 mg/l really not being justifiable. He said that he applauded the way the NFU had managed to lobby Parliament on a whole series of matters, but in this instance it was European legislation. He believed that the NFU would find it more difficult to change the minds of those in Brussels.

He felt that the water industry would be delighted if the limit could be raised because then they would not have the problems of treating the water that was currently above the limit. He invited Mr Payne to comment on how a change in the law could be achieved.

Authors' replies to discussion on Papers 4, 5, 6, and 7

PAPER 4

Mr H. G. Headworth, replying to Mr Allcock, confirmed that Southern Water had an aquifer protection policy developed from 1976. It was in keeping with those which other authorities had. He said that aquifer protection policies were very good for point sources of pollution, but were less useful for diffuse sources of pollution. In the case of the Southern Water policy, the aquifers had been considered from the point of view of those which were most vulnerable and those which were least vulnerable, with the areas around public sources obviously being at greatest risk. For each of these zones, activities were matched against them which it was desirable to prohibit or control in order to reduce or eliminate the pollution risk. This covered landfill, storage of materials, drilling, oil exploration, and where possible, point sources of farming waste.

He felt that the problem lay with diffuse sources of pollution, such as pesticides, using sprays, and nitrates. It was a problem which all European countries recognized. He said that the merit of an aquifer protection policy was that it could be constructive in offering alternative areas for use, but this could obviously not apply to diffuse sources of pollution, and it was difficult to know where the solution lay.

Mr Headworth said that another aspect of aquifer protection policies was that they were essentially consultative rather than statutory. The statutory powers that could be made use of were quite limited. Their success lay in the fact that local authorities, industrialists, and farmers, in some cases, knew of the policy and were able to discuss with the water authority remedies or changes in practice that should be undertaken to lessen the risk of pollution.

With regard to sheep-dipping he said that in his area sheep-dipping was carried out by itinerant sheep-dippers, but the water quality officers had maintained good liaison and discussed with them how best to overcome the disposal of waste.

Mr Headworth said that he was not sure that he was really qualified to answer the point on the significance of the concentration of pesticides. The concentrations that he had given in the paper were very low, and, in his view, the problem in terms of ground-water in the UK should not be overstated.

He said that the increase in nitrate usage, as illustrated by Mr Allcock, was indisputable and had led to the problem which he had tried to illustrate. He said that Southern Water, whilst recognizing that there was little that could be done in the areas of the Chalk, where the 40–50 m of unsaturated chalk meant a 20–40 year time lapse before any action taken at the surface could be seen to have any counter action at the groundwater table, realized that that did not mean that action should not be taken. Regular meetings with farmers on Thanet, which was the only part of the Southern Water area where there was a nitrate problem, had been held to discuss the problem with them to understand the Authority's viewpoint, and to discuss what action they might take to lessen the use of fertilizer and control the timing of its use. He felt that these meetings had been constructive, although he was not sure that it had changed things a great deal as yet.

In answer to Mr Aspinwall with regard to pesticides, he said that in the Southern Water area he did not consider that there was a problem. He said that the higher concentrations in the area occurred in a few sources in Kent, whereas current monitoring in the rest of the area indicated values less than the detection limit. He said that the higher values tended to occur more in urban areas, and felt that this supported the view that it was more likely to result from the spraying alongside roads and railways than from use in agriculture.

With regard to the status of the sampling, he said that the results which he showed were derived from pumped sources. He knew of no work that had been done on the unsaturated zone for pesticide concentrations.

He said that he would not wish to speculate on whether the patterns one might see were likely to parallel those profiles related to nitrates.

PAPER 5

Mr L. Beck agreed with Mr Allcock that external peripheral drains for silage clamps were desirable. He had heard this recommended at a number of recent farm waste conferences, and it was certainly a recommendation he would endorse.

With regard to leakage through floors, he said that there was currently a move toward the use of hot-rolled asphalt, and this might prove to be safer than concrete.

Mr Beck endorsed Mr Black's plea for a greater degree of discussion between Government, the authorities, and the farmers, especially where, as in Mr Black's case, much had been done to solve problems, only to find that it was still not enough.

In response to Mr Aspinwall, he said that, on the question of whether or not there were problems of pesticide residues in ground and surface waters, he thought there might be groundwater problems in certain parts of the UK, but so far as Yorkshire was concerned, the problems tended to arise on surface waters, and these were usually one-off incidents or a short series of one-off incidents.

Mr Beck explained that an investigation was being carried out into the disposal of sheep-dip wastes on the Yorkshire Wolds. There had been concern about the practice of using soakaways dug into the chalk. There were itinerant contract sheep-dipping units, and it was not always known what they did with the spent dip. A monitoring programme had been set up on the groundwaters. To date the analysis results were looking hopeful.

He agreed with Mrs Warr that there was a need for environmental awareness to be built into agricultural college courses, and he hoped that water authority officers, NRA officers (or even retired river quality managers!) could be enlisted to assist with lectures if necessary.

Responding to Mrs Warr's question about commercial pressures on farmers to buy additives etc., Mr Beck said his experience of most reputable companies was that they did give correct information about whether or not silage additive would assist in reducing effluent production, but he knew of at least one company which had tried to claim that if their product was used, the farmer would not need to worry about effluent, and this was to be deplored.

In answer to a question about the effectiveness of grant aid, Mr Beck said that many of the present problems could be dealt with through grant aid because the problem required the provision of new buildings and plant, but in both the old and new MAFF schemes there was no provision for assistance with the separation of surface waters, in spite of protestations from the WAA. Although it was now too late to change the details of the new scheme introduced this year, it might be a good idea if this Institution could add its weight to the pressure already being applied to MAFF by the regulatory authorities to get this omission put right at an early opportunity.

He endorsed Mr Stevens' comment that high quality materials should be used. He said that one of the points made in his paper about things going wrong on the farm was the high frequency of disasters caused by poor quality workmanship.

In response to a written question by Mr Burroughes about the Friends of the Earth Report on UK potable supplies which had exceeded the EC pesticide limit, Mr Beck confirmed that the water authorities were indeed aware of the report's contents, since they had themselves provided the data for raw water sources in response to a request by Friends of the Earth.

Incidents were caused in both groundwaters and surface waters, and although some could be described as diffuse pollution, in many cases they could still be related to a known specific activity.

Mr Burroughes had also asked about over-stocking in relation to available land being a contributing factor in slurry pollution incidents, and Mr Beck confirmed that there were many such cases. For example, in the pig-rearing areas of North Humberside the 'answer' in recent years to the ever-increasing volumes of slurry had been the progressive construction of huge earth-banked long-term storage lagoons. Breaches in embankments often occurred, and sometimes these were about the same size and shape as a spade!

PAPER 6

Mr Payne, in response to Mr Binnie, admitted that to achieve a change in the nitrate standard would pose some problems. He agreed that in the present political climate in Europe it was not possible to get that limit reduced. However, the NFU drew attention to the lack of scientific evidence supporting the EC standard, particularly bearing in mind the implications of compounding past errors in future legislation. A draft Directive had been published during December 1988 proposing further controls on nitrate levels aimed at ensuring that the problem was dealt with at source—a superficially attractive proposition which posed major problems in the context of nitrates.

In answer to Mr Aspinwall's suggestion that the NFU was negative to organic farming, Mr Payne said that he did not accept that. The NFU had a member of staff who was specifically responsible for organic farming, and he did not think that those organizations responsible for organic farming would support the view that the NFU was negative about it.

With regard to his specific comments, Mr Payne said that they were intended to reflect the evidence as the NFU understood it. As he had indicated, the worst farming operation from the point of view of releasing nitrate was ploughing grassland. This was an inherent part of the organic system, and the NFU had actually discussed the question of nitrate leaching with the Elm Farm Research Centre, whose preliminary research indicated quantities that would typically leach from an organic farm (at least in areas of the eastern counties where there was low rainfall) causing the European nitrate standard to be exceeded.

Referring to the organic content of soil, he said that he was aware that in recent years there had been a good deal of concern that the switch from the traditional systems of rotational farming to continuous arable cropping was leading to a reduction in organic content, but the understanding from soil scientists was that organic matter content in arable soils was now increasing and had been for some years.

In reply to Mr Turley, he said that he did not feel qualified to answer whether injection would lead to helping with the nitrate problem.

Turning to pollution from farm waste, he said that Mr Merriman had made a very constructive suggestion of looking at a scheme specifically aimed at dairy farmers. He considered that there was a good case for this, as also there probably was for including the separation of clean water from dirty water under the grant scheme. He said that the NFU would certainly be very willing to discuss possibilities with the water authorities for new initiatives in this area.

In reply to Mr Braithwaite, he said that grant aid of 50% on those items which were eligible would clearly help, although he echoed comments of other contributors concerning the inadequacy of the scheme. He was much concerned about those who could not afford to take the action which would be required of them. It had been correctly stated that the problems stemmed from intensification in the agricultural industry, particularly the dairy industry. That had been a matter of government policy which had been implemented by means of excellent advice from ADAS over the years and by reducing real prices, so that it had been necessary to produce a greater output to maintain farmers' income. However, the question must now be asked about society's objectives, whether or not catchments were able to sustain the intensity of dairy use which we now had and which farmers had moved toward with full support from the Government's own advisory service, and whether or not this was consistent with environmental objectives. He said that there was a need to step back and to reconcile agricultural and environmental policies for the future.

Referring to the achievement of a culture change, Mr Payne said that he thought that Mr Booth was correct, and one of the most powerful pressures that could be exerted on farmers was peer pressure from their neighbours. He felt that this had been particularly effective in the case of straw burning. The agricultural industry's image had been suffering from the actions of an irresponsible minority. The NFU produced its own code of practice with the support of all the relevant organizations, and sought to enforce that code largely through the medium of peer pressure, together with back-up from byelaws. He knew of many cases where there had been problems with individual farmers not obeying the code where their neighbours had 'brought them into line'.

He felt that the culture change could be achieved partly through the process of education and partly through a system of peer pressure, and to achieve the latter one would need to bring home to farmers that there would be unwelcome consequences if agriculture did not improve its performance. He said that this was exactly the message that the NFU was seeking to get over in its pollution campaign. The leaflet was being circulated to members and was being sent to all dairy farmers through the Milk Marketing Board. Any help from other sources would be very much welcomed by the NFU.

PAPER 7

Dr Schofield said that she thought that there was still a general lack of knowledge amongst farmers as to exactly how they were affecting river systems, especially with respect to parlour and yard washings. She felt that there was still a great need for on-farm visits from water authorities and others. Many water authorities had conducted campaigns to point out where farmers were likely to have potential gross failures, such as slurry store and silage liquor leaks, and in most cases farmers had taken heed of this. However, there was still a problem with chronic pollution, and she thought that much awareness was still required concerning what needed to be done with the general farmyard wastes. That was very intensive in respect of staff time, and water authority staff could not afford the time that was required to do it effectively.

Dr Schofield said that one of the objectives of the study that had been undertaken was to try to aid the farming community and to come up with more cost-effective methods of wastewater treatment. At present a review of methods was being carried out. Having talked to the farmers involved, she said that the application of reed-bed systems would be acceptable to them. However, she agreed that from the point of view of amm.N treatment the results might not indicate that the process would be ideal, but very few tests had been carried out, and the whole point of trying to do it on an on-farm basis was that many of them were very experimental approaches using domestic sewages.

She accepted that the wastewaters were very high in amm.N content (up to 140 mg/l) and that that might expect a lot of a reed-bed system.

In response to Dr Montgomery, she said that buffer zones had not been perceived as being an effective solution to yard water washings. They were seen as being of use on land bordering streams and acting as a boundary zone between land being spread with slurry and the stream.

Quite a lot of farms had such areas naturally—reed, marsh, or coppice areas adjacent to the streams. It was simply intended to observe whether they were effective in preventing land run-off and slurry spread on the land from polluting streams.

In answer to Mr Oxley, she said that the area in which the work had been carried out was not limestone, but shales and mudstones. From her point of view the geology of the area was taken into consideration. However, she doubted if this was the case from the farmers' point of view. She accepted that certain areas from a more global point of view would not be any good as buffer zones. One of the problems that that part of the project faced was that, because of the shale type of bedrock, there was a potential that even if the amm.N leached through the soil, which was quite shallow in places, then one would get natural boreholes in the bedrock and run-through straight through the slate and shale. It was hoped to monitor this.

One of the main points of the research was to try to monitor how much pollution was coming off the farms and how much was coming off the land, in order to facilitate the decision as to where to aim the solutions. It was very difficult to try to quantify what was coming off the land.

Chapter 3

Water quality criteria for water use in the agricultural sector causes a great deal of concern to the regulatory authorities. The extensive use that is made of our river systems as transportation media for treated wastes, land run-off, and as conveyancers of storm water result in wide variations in quality. The contributions in this chapter explore the agricultural requirements for water with particular reference to irrigation and fish farming.

Fish farming is viewed as an agricultural activity, and can cause major pollution of river systems. Fish farms continue to grow in the numbers and size and cause concern to other agricultural users of water because of their location in areas of high water quality. Jeopardizing water quality could well affect the ecology of the river and the ability of that river to support salmonid fisheries. Water resources protection and the concept of protection zones are discussed in the final paper of this chapter. The concept of nitrate sensitive areas and the reasons from departing from the 'polluter pays' principle is related to the proposed European Directive on the control of nitrates. Similarities between the UK and the European position are illustrated.

8

The agricultural requirement for water, with particular reference to irrigation

R. J. Bailey, BSc, PhD [†] and J. Minhinick, BA [‡]

ABSTRACT

A quantitative analysis of irrigation requirement in different climatic areas of England and Wales is presented. The analysis takes account of the differing requirements of various crops and variation in soil texture. The results include an estimation of irrigation requirement in an average year, a typical dry year for irrigation planning purposes, and an exceptional year such as 1976. The analysis was conducted by using a computer program developed jointly by ADAS and the Meteorological Office, and a brief description of the salient features of the program is given.

KEY WORDS Irrigation   Crop requirements   Licensing   Soil moisture deficit   Climatic variation

INTRODUCTION

A farm requires water for various purposes which, apart from domestic use for the family, may include crop spraying, vegetable washing before packing, drinking for livestock, washing dairy buildings, dilution and handling of farm effluent, and irrigation. This paper is concerned with the requirement for irrigation, but it is as well to place these other uses in perspective. The dairy industry is the biggest agricultural user of water. The drinking and washing requirements amount to 150 litres per day for each cow. Taking the size of the national herd as 2.5 million cows, this represents a total requirement for 135 million cubic metres per annum. This requirement is fairly evenly spread throughout the year. The requirements for other sectors of farming are considerably smaller. These figures are already published [1, 2], and no further discussion of them is necessary here.

In contrast, the requirement for irrigation of crops is variable. For instance, two recent MAFF surveys have shown that 55 million cubic metres were used in 1982, and 98 million in 1984. Also, irrigation does not represent a constant demand, but is restricted to the summer months and, even during this period, shows major fluctuations with peaks

---

[†] National Irrigation Specialist, ADAS, Ministry of Agriculture, Fisheries and Foods.
[‡] Regional Agricultural Meteorologist, Meteorological Office, on attachment to ADAS, Ministry of Fisheries and Food.

corresponding to periods of low rainfall. The purpose of this paper is to describe the irrigation requirements of various crops in different part of England and Wales, both in a 'typical' year, and in an extreme year such as 1976.

METHODS

The ADAS/Meteorological Office computer program

A computer program to estimate irrigation requirement was developed jointly by the Agricultural Development and Advisory Service (ADAS) and the Meteorological Office. The required inputs consist of soil texture, the appropriate irrigation plan, and certain agronomic data for each crop under consideration, for example, a typical date of planting, typical dates of attaining 25% and 75% crop cover, etc. The program has access to weather data for 1961–1980 from several hundred weather stations based on a key network of 94 weather stations in England and Wales that have complete records. When calculating irrigation requirement for an individual farm, the nearest relevant weather station is selected, and the weather from each year is analyzed. A balance sheet approach is used, with rainfall as a credit, and evapotranspiration as a debit. Evapotranspiration is calculated by using a modified form of the Penman–Monteith combination equation [3]. During the analysis, a record is logged of every occasion on which irrigation would have been required to prevent a crop from becoming stressed by lack of moisture, and the subsequent balance sheet assumes that this irrigation has been applied. At the end of the analysis, the amount and dates of irrigation required for each crop in each year are produced.

For the present purpose, where an overall view of irrigation requirement is desired, rather than that of a specific farm, it was necessary to modify the approach. It was not practical to run the program for every holding in England and Wales, but farms were placed into categories of climate and soil type, so that a set of data obtained from each category could be extrapolated to a large number of farms. This technique was considered to be sufficient for the calculation of average irrigation requirement over large areas, but it would be erroneous to take the figures presented here out of context and use them as a basis for irrigation planning on a specific farm.

Climatic variation between years

The climatic variation that exists between years is such that an analysis of this nature could be expected to produce a variable irrigation requirement over the years 1961–1980. Following convention [4] the fifth driest year was selected as that representing a suitable weather pattern for irrigation planning. This fifth driest year will be subsequently referred to as a 'typical' year. Also, so that the requirement in an extreme year can be appreciated, the data from 1976 are also reported.

Climatic variation between localities

The climate in the UK is such that western districts receive much more rainfall than the eastern side of the country. Irrigation requirement is also dependent on the amount of sunshine that a crop receives, and this varies with locality, the highest amounts being found in southern districts.

The influence of climate on irrigation requirement can be appreciated by an examination of the Soil Moisture Deficit (SMD) attained in each area. The Meteorological Office have classified the whole of England and Wales into 52 climatic areas [5]. One of the parameters published for each area is a median value for the maximum SMD attained in each year, based on 1941–70. This information was used in the present analysis to place each of the 52 areas into one of seven categories. The grouping is shown in Fig. 8.1; at boundaries, if there are non-adjacent categories, there will be a small transition zone not shown at this resolution. For each category, a representative weather station was then selected to provide data for the calculation of irrigation requirement, except for area 1 which is not represented by a weather station. The selected stations are shown in Table 8.1.

Table 8.1. Selected weather stations representative of each climatic area

Climatic area	Median value of max SMD (mm) (1941–70)	Representative weather station	Median value of max SMD (mm) (1961–80)
1	0–25	None	
2	26–50	Cwmystwyth, Dyfed	37
3	51–60	Helmshore, Lancashire	55
4	61–85	Silpho Moor, Yorkshire	79
5	86–95	Newport, Shropshire	91
6	96–105	Fernhurst, W. Sussex	97
7	106–120	Wattisham, Suffolk	111

Effect of soil type on irrigation requirement

Differences in soil provide another source of variation in irrigation requirement. During periods of low rainfall, a crop will extract all or part of its requirement from the available soil reserve, but this varies with soil texture. In the present analysis, four soils were chosen to represent the range of soil types likely to be found. These soils, together with their calculated available water capacities, are shown in Table 8.2. Within each of the seven climatic groups, the irrigation requirement was calculated for each soil type.

Table 8.2. The four representative soil types used in the analysis, and their respective available water capacities (derived from Hollis [6])

Texture class		Topsoil available water % (Easily available %)		Subsoil available water % (Easily available %)	
Topsoil (30 cm)	Subsoil				
Loamy coarse sand	Coarse sand	11	(7)	7	(5)
Medium sandy loam	Medium sandy loam	17	(11)	15	(11)
Silt loam	Silt loam	23	(15)	22	(14)
Loamy peat	Loamy peat	35	(26)	35	(26)

Irrigation requirement and type of crop

To establish an irrigation plan, it was necessary to take account of the rooting depth commonly associated with each crop. The rooting depths used in the present analysis are presented in Table 8.3. The data in Tables 8.2 and 8.3 were combined and used to

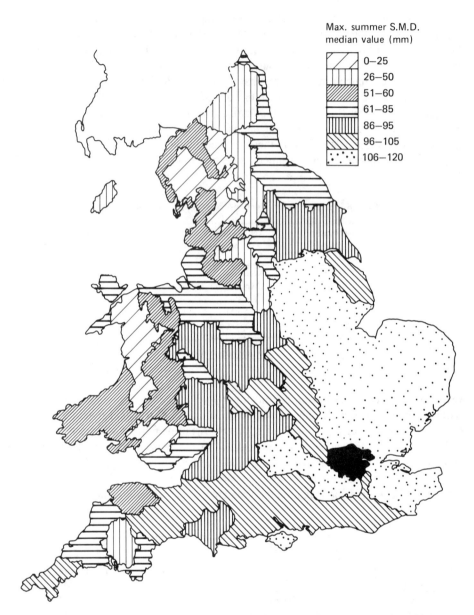

Fig. 8.1. Agroclimatic areas of England and Wales, based on maximum summer Soil Moisture Deficits. See Table 8.1.

calculate the root zone capacity for each crop in each soil type; that is, that part of the available water found within the rooting zone. For this purpose, the model of Hall *et al.* [7] was used; that is, it was assumed that a crop may extract water up to a tension of 15 bar (available water) throughout the zone in which roots proliferate, but in the zone where roots are more sparse, the amount of water held to 2 bar (easily available water) was taken as being representative of the amount of water available to the crop.

Table 8.3. Rooting patterns of various crops

Crop		Depth in cm of intensive rooting(capable of extracting available water, i.e., up to 15 bar)	Depth in cm of extensive rooting (capable of extracting easily available water only, i.e., up to 2 bar)
Early potatoes	*	50	60
Maincrop potatoes	**	70	–
Sugar beet: in June	***	25	50
in July	***	50	100
in August	***	80	130
in September	***	100	160
Cereals	**	50	120
Grass (permanent)	**	100	–

Source: *   Aslyng [8]
        **   Hall *et al.* [7]
        *** Dunham [9, 10]

The next step was to calculate a critical deficit (that is, that deficit at which yield or quality become affected) for each combination of crop and soil type. The precise relationship between critical deficit and available water is very complex, and varies with crop, soil, and prevailing weather conditions. However, Aslyng [8] has reported that it is sufficiently accurate for irrigation planning purposes to assume that production is limited by lack of water when more than 50–60% of the root zone capacity has been exhausted. In this instance, the critical deficit was therefore simply estimated by halving the root zone capacity.

The critical deficits are presented in Table 8.4.

Table 8.4. Estimated critical soil moisture deficits for various crops on a range of soil types

Soil	Critical Deficits (mm)							
	Early potatoes	Maincrop potatoes	Sugar beet				Cereals	Grass
			June	July	Aug	Sep.		
Loamy coarse sand over coarse sand	26	30	20	36	46	56	41	44
Medium sandy loam	46	55	35	68	90	111	79	78
Silt loam	63	78	46	91	124	153	105	111
Loamy peat	100	122	76	152	205	253	178	143

The irrigation plans used for each crop were designed to irrigate with 25 mm water every time that the SMD reached the critical deficit, except that only 20 mm was used for sugar beet on a loamy coarse sand in June. An additional complication with maincrop potatoes is that many growers are irrigating to control the level of common scab disease

(*Streptomyces scabies*). This requires very intensive irrigation (12 mm water whenever SMD reaches 15 mm) during the six weeks immediately after tuber initiation, and this regime was included in the irrigation plans used for this crop.

RESULTS AND DISCUSSION

The results are presented in Tables 8.5 to 8.7. From these tables, it is possible to obtain an approximate irrigation requirement for a range of crops, grown in a range of soils, in each of the climatic areas 2–7. It can be safely assumed that there is no irrigation requirement in area 1. It must be emphasized that the data represent a biological requirement for water, and do not imply an economic level of irrigation requirement. The economics are continually changing, being dependent on the value of the crop and the cost of irrigation, and they have not been taken into account in this analysis. As an example, until recently it has been an economic proposition to irrigate cereals on sandy soils in climatic area 7. However, in recent years the cost of irrigation has increased without a corresponding increase in the price of cereals, and irrigation of this crop has become difficult to justify.

Table 8.5. Median irrigation requirement (10th driest year in 20) for various crops

Climatic area	Soil type	Median irrigation requirement (mm)				
		Early potatoes	Maincrop potatoes	Sugar beet	Cereals	Grass
2	Loamy coarse sand	0	25	25	0	25
	Medium sandy loam	0	25	0	0	0
	Silt loam	0	25	0	0	0
	Loamy peat	0	25	0	0	0
3	Loamy coarse sand	0	50	25	0	50
	Medium sandy loam	0	50	0	0	0
	Silt loam	0	50	0	0	0
	Loamy peat	0	50	0	0	0
4	Loamy coarse sand	0	100	50	0	100
	Medium sandy loam	0	75	0	0	50
	Silt loam	0	75	0	0	25
	Loamy peat	0	75	0	0	0
5	Loamy coarse sand	25	100	75	25	125
	Medium sandy loam	0	75	0	0	75
	Silt loam	0	75	0	0	50
	Loamy peat	0	75	0	0	25
6	Loamy coarse sand	25	125	75	25	125
	Medium sandy loam	0	75	0	0	100
	Silt loam	0	75	0	0	50
	Loamy peat	0	75	0	0	25
7	Loamy coarse sand	50	125	125	25	175
	Medium sandy loam	0	75	25	0	125
	Silt loam	0	75	0	0	100
	Loamy peat	0	75	0	0	75

Table 8.6. Irrigation requirement for planning purposes (5th driest year in 20) for various crops

Climatic area	Soil type	Planning irrigation requirement (mm)				
		Early potatoes	Maincrop potatoes	Sugar beet	Cereals	Grass
2	Loamy coarse sand	25	50	25	25	50
	Medium sandy loam	0	50	0	0	0
	Silt loam	0	50	0	0	0
	Loamy peat	0	50	0	0	0
3	Loamy coarse sand	25	75	50	25	75
	Medium sandy loam	0	50	0	0	25
	Silt loam	0	50	0	0	0
	Loamy peat	0	50	0	0	0
4	Loamy coarse sand	25	125	75	25	125
	Medium sandy loam	0	100	25	0	100
	Silt loam	0	100	0	0	50
	Loamy peat	0	100	0	0	25
5	Loamy coarse sand	50	150	100	50	175
	Medium sandy loam	25	100	25	0	100
	Silt loam	0	100	0	0	75
	Loamy peat	0	100	0	0	50
6	Loamy coarse sand	50	150	100	50	175
	Medium sandy loam	25	100	25	0	125
	Silt loam	0	100	0	0	75
	Loamy peat	0	100	0	0	50
7	Loamy coarse sand	50	175	150	50	250
	Medium sandy loam	0	125	75	25	200
	Silt loam	0	100	25	0	175
	Loamy peat	0	100	0	0	150

Table 8.7. Maximum irrigation requirement (driest year in 20) for various crops

Climatic area	Soil type	Maximum irrigation requirement (mm)				
		Early potatoes	Maincrop potatoes	Sugar beet	Cereals	Grass
2	Loamy coarse sand	50	175	100	50	150
	Medium sandy loam	25	100	50	25	125
	Silt loam	0	100	0	0	100
	Loamy peat	0	100	0	0	50
3	Loamy coarse sand	50	175	125	100	200
	Medium sandy loam	25	125	50	50	150
	Silt loam	0	100	25	25	125
	Loamy peat	0	100	0	0	100
4	Loamy coarse sand	50	225	175	50	250
	Medium sandy loam	25	175	100	25	225
	Silt loam	0	125	50	0	200
	Loamy peat	0	150	0	0	150
5	Loamy coarse sand	75	275	200	125	350
	Medium sandy loam	25	200	125	75	300
	Silt loam	25	125	100	50	275
	Loamy peat	0	175	0	0	250
6	Loamy coarse sand	75	300	225	125	425
	Medium sandy loam	25	225	125	75	400
	Silt loam	25	175	100	50	350
	Loamy peat	0	200	0	0	325
7	Loamy coarse sand	75	300	225	125	425
	Medium sandy loam	25	250	125	75	400
	Silt loam	25	200	100	50	350
	Loamy peat	0	200	25	0	325

The irrigation requirement of maincrop potatoes in the wetter climatic areas does, initially, appear to be rather high, but this is not surprising. The irrigation plan used here is designed to produce high yield and high quality. As explained previously, high quality means controlling common scab, which requires intensive irrigation. There is much experimental evidence [11, 12, 13] to show that applying 12 mm irrigation whenever the SMD approaches 15 mm for a six week period starting at tuber initiation, is an effective method of controlling the disease on light soils. There is no published information regarding an irrigation regime to use on medium and heavier soils, although the disease is commonly found. Until some information becomes available, farmers are advised to use 12 mm at 15 mm SMD on all soil types, and this regime has been adopted in this analysis for all soil types. Even wet climatic areas within the UK typically experience short periods of dry weather in June, allowing the SMD to exceed 15 mm and thereby requiring the levels of irrigation shown here.

Finally, the requirements arrived at in this analysis are far in excess of the amounts being used at present by most farmers. There are various reasons for this. It may be due in part to a lack of machinery or labour. It may also be due to the fact that many farmers practise no formal irrigation scheduling, and are unaware of the water requirement of their crops, but apply irrigation according to rather crude criteria. This situation is changing, and, with more awareness, greater amounts of irrigation water may be applied in future dry years. The Strutt report [2] predicted in 1980 that the irrigation demand would quadruple in 20 years. There has been little evidence of this trend, mainly owing to a run of wet summers recently. However, if the demand for quality, and need for reliably high yields, affect the economic appraisal of irrigation in future years such that the demand does increase, the analysis presented here should serve as a reliable guide to depth of irrigation required in a typical year, a dry year, and an exceptionally dry year.

The 1984 MAFF survey presented data for the irrigated area of each crop in England and Wales, totalling 140 620 ha. A reasonable estimate of total irrigation requirement can probably be obtained by applying the amounts of irrigation in Tables 8.5 to 8.7 over the area irrigated in 1984, assuming a loamy coarse sand for early potatoes, and a medium sandy loam for other crops. Vegetables can also be taken into account by assuming similar amounts to those required by potatoes. The results are presented in Tables 8.8 to 8.10.

Table 8.8. Median irrigation requirement ('000 m^3)

Climatic area	Potatoes Early	Maincrop	Sugar beet	Cereals	Grass	Veg	Total
1	0	0	0	0	0	0	0
2	0	85	0	0	0	90	175
3	0	220	0	0	0	370	590
4	0	2 790	0	0	985	315	4 090
5	450	5 100	0	0	3 595	2 065	11 210
6	105	1 285	0	0	3 790	810	5 990
7	1 990	16 200	5 025	0	8 690	9 060	40 965
	2 545	25 680	5 025	0	17 060	12 710	63 020

Table 8.9. Fifth driest year in 20 irrigation requirement ('000 m³)

Climatic area	Potatoes Early	Maincrop	Sugar beet	Cereals	Grass	Veg	Total
1	0	0	0	0	0	0	0
2	10	170	0	0	0	175	355
3	225	220	0	0	245	370	1 090
4	105	3 720	265	0	1 970	420	6 480
5	900	6 800	1 060	0	4 790	2 750	16 300
6	210	1 710	25	0	4 740	1 080	7 765
7	1 990	27 000	15 075	4 520	13 900	15 100	77 585
	3 470	39 620	16 425	4 520	25 645	19 895	109 575

Table 8.10. Exceptional dry year irrigation requirement ('000 m³)

Climatic area	Potatoes Early	Maincrop	Sugar beet	Cereals	Grass	Veg	Total
1	0	0	0	0	0	0	0
2	20	340	0	30	540	350	1 280
3	510	555	10	50	1 455	925	3 505
4	210	6 510	1 060	450	4 430	735	13 350
5	1 350	13 600	5 285	3 090	14 370	5 500	43 195
6	315	3 850	115	475	15 160	2 430	22 345
7	2 985	54 000	25 125	13 560	27 800	30 200	153 670
	5 390	78 855	31 595	17 610	63 755	40 140	237 345

In 1988, irrigation licences in England and Wales amounted to over 200 million cubic metres. The present analysis indicates that this may well be justified in an exceptional year, but is far in excess of that required in an average year (63 million) or even the fifth driest year in twenty (110 million). It is possible that the area irrigated in 1984 does not represent the maximum area that farmers would irrigate in a dry year, and the amounts of irrigation estimated here should be applied over a larger area. Indeed, in the 1984 MAFF survey, farmers were asked to submit details of the area that they would probably irrigate in a very dry year, and these amounted to 189 000 ha, an increase of 35%. Given the current situation with regard to labour and irrigation equipment found on most farms, however, it is unlikely that the full required rate of irrigation could be sustained over this larger area in a dry year.

Although the approach adopted here will inevitably contain some inaccuracies, they are unlikely to account for the fact that existing licences are double the required quantity estimated here for the fifth driest year in twenty. The analysis presented here suggests that existing licences are not closely related to needs, and the evidence indicates that a more intensive study is required in some areas.

REFERENCES

[1]   Anon (1980) Farm Water Supply—Emergency Planning. *ADAS leaflet 629.*

[2]   Strutt, N., *et al.* (1980) Water for agriculture: Future needs. *Report of the Advisory Council for Agriculture and Horticulture in England and Wales.*

[3]   Thompson, N., Barrie, I. A. & Ayles, M. (1981) The Meteorological Office rainfall

and evaporation calculation system: MORECS. *Meteorological Office Hydrological Memorandum No. 45.*

[4]   Anon (1982) Irrigation. *MAFF/ADAS Reference Book 138.*

[5]   Smith, L. P. (1984) The agricultural climate of England and Wales. *MAFF/ADAS Reference Book 435.*

[6]   Hollis, J. M. (1987) The calculation of crop adjusted soil availability water capacity (AP) for wheat and potatoes. *Soil Survey Research Report No. 87/1.*

[7]   Hall, D. G. M., Reeve, M. J., Thomasson, A. J. & Wright, V. F. (1977) Water retention, porosity and density of field soils. *Soil Survey Technical Monograph No. 9.*

[8]   Aslyng, H. C. (1984) Soil water capacity, climate and plant production. The development and future of irrigation. In: *Proceedings of the North-Western European Irrigation Conference*, Billund, Denmark, 2–39.

[9]   Dunham, R. J. (1987) Irrigating sugar beet in the United Kingdom. In: *Proceedings of the 2nd Northwest European Irrigation Conference*, Silsoe, England (In press).

[10] Dunham, R. J. (1988) *Personal Communication.*

[11] Lapwood, D. H., Wellings, L. W. & Rosser, W. R. (1970) The control of common scab of potatoes by irrigation. *Annals of Applied Biology*, **66**, 397–405.

[12] Lapwood, D. H., Wellings, L. W. & Hawkins, J. H. (1971) Irrigation as a practical means to control potato common scab (*Streptomyces scabies*). *Plant Pathology*, **20**, 157–163.

[13] Lapwood, D. H., Wellings, L. W. & Hawkins, J. H. (1973) Irrigation as a practical means to control potato common scab (*Streptomyces scabies*); final experiment and conclusion. *Plant Pathology*, **22**, 35–41.

9

Water requirements for fish farming

C. G. Saunders-Davies [†]

ABSTRACT

Fish farming has been practised for thousands of years all over the world. The Far East has a very substantial production and China is by far the World's largest fish farming nation with an annual production in excess of two million tons per year.

Fish farming takes many forms with cultivation of warm water species such as carp and tilapia often practised on a traditional extensive basis; but also includes the cultivation of many marine species, including shell fish and crustaceans, as well as salmonid cultivation, which is practised in Northern and Western Europe and which is the type of fish farming which most readily comes to our minds here.

The water quality needs of fish farming are reviewed as is the impact of discharges from fish farms on receiving waters.

KEY WORDS   Fish farm   Water quality discharges   Water quality

INTRODUCTION

Fish farming has been practised for thousands of years all over the world. The Far East has a very substantial production, and China is by far the World's largest fish farming nation, with an annual production in excess of 2 million tons per year.

Fish farming takes many forms with cultivation of warm water species such as carp and tilapia often practised on a traditional extensive basis; but it also includes the cultivation of many marine species, including shell fish and crustaceans, as well as salmonid cultivation, which is practised in Northern and Western Europe and which is the type of fish farming which most readily comes to our minds.

Thus the Western European production of around 200 000 tons of salmonids can be seen in the worldwide context of a production of many million tons of a wide variety of fish species.

Table 9.1 shows the growth of trout and salmon production in the UK in recent years.

It will be seen that trout production has grown rapidly over the last 15 years, but this

[†] Test Valley Trout Ltd.

rate of expansion is likely to slow, limited by resources, whereas salmon production has taken longer to get into its stride owing to the need to solve greater technical problems, but is now set for a period of substantial expansion, especially with the use of more exposed marine sites, made possible by improved technology.

Table 9.1. Trout and salmon production in the United Kingdom

	Trout	Salmon
1975	1 300	
1976	1 300	
1977	2 000	
1978	3 100	
1979	4 500	520
1980	5 500	598
1981	6 500	1 133
1982	6 000	2 152
1983	7 000	2 536
1984	8 500	3 912
1985	9 500	5 921
1986	11 000	10 337
1987	14 000	12 721
1988 (est.)	15 000	22 000
1989 (est.)	16 000	30 000

In this paper I shall concentrate on salmonid culture. Trout farming is practised mainly in fresh water conditions (although there is some production of large trout in salt water). In England and Wales this is predominantly in stream fed farms, but in Scotland there has been a considerable growth in the use of cages in fresh water lochs.

The fattening phase for Atlantic salmon requires marine conditions, and most salmon farming within the UK takes place off the west coast of Scotland and in the western and northern isles. The production of smolts for salmon farming requires fresh water, and again it is predominantly carried out in Scotland. Originally, smolt farms were based on flowing water, but recently smolts are being produced most successfully in fresh water cages, and this has been a factor enabling smolt production to increase substantially to meet a rapidly growing demand.

WATER QUALITY NEEDS

Salmonid farming requires a very high quality water, and it is for this reason that fish farms have become established on many high quality rivers, having DoE classification of class 1. Fish farms are established throughout the British Isles, and are inevitably found in greatest numbers on the high quality fishing rivers which, in providing suitable conditions for their natural salmonid populations, are therefore well suited for the needs of fish farming.

As well as quality, temperature and flow rates are of vital importance. Some very large trout farms have been established in continental Europe and the USA on large artesian water sources having constant temperatures ranging from 10–14°C. These conditions do not exist in the UK, but farms where the temperature averages 10°C or more over the year and lies between 6 and 16°C would give excellent performance. Higher temperature can give rise to problems, and fish cannot be fed. Ideally, the temperature should never go above 20°C.

The generally accepted figure for production without the use of recirculation or re-aeration is that a flow of 5 litres per second is required for every annual ton of production. Since it is the minimum rather than the average flow which determines the available water supply, especially as this minimum normally occurs at the time of the greatest need, spring fed rivers with a relatively stable flow provide the best type of water supply. Rivers suffering very wide extremes of flow often result in water quality problems with very high temperatures at low flows and a very high suspended matter in spate conditions.

In cage farming, water quality is equally important, but changes take place at a much lower rate, and the water environment is itself likely to be affected by the existence of the fish farm. This has led to the requirement in some cases for an environmental assessment to be made before any developments are undertaken. I would refer particularly to the work done by Beveridge, Philips, and Ross at Stirling University over the last few years, in quantifying some of these factors.

In running water it is possible to consider the treatment of effluents to improve their quality before return to the river system. This option is not available to still water and marine sites. Coupled to the treatment of fish farm effluents is the concept of re-use or aeration of the water to achieve greater production from any given flow. Although this has many theoretical attractions, its use in salmonid culture is not widespread, mainly for economic reasons, although the complexity and the fact that a life support system is dependent on mechanical and electrical components, are also factors. In some cases limited recirculation and/or re-aeration of water to overcome temporary low flow is used successfully.

FISH FARMING IN THE CONTEXT OF OVERALL WATER USE

Since the early 1970s aquaculture in the UK has grown from a small beginning to an important sector of agriculture production which in 1989 will be worth more than £100 000 000 at first sale value. Accompanying this growth in output there has been an increasing public awareness of the industry.

As fish farming production rises in the UK the bulk of the increase will come from salt water cage production, and it is unlikely that production in fresh water will be able to rise at anything like the rate that it has achieved in the last decade, owing to there being a limitation on the availability of suitable new sites.

No separate figures are produced for production divided between still water and flowing water, although it can be assumed that most production in England and Wales uses flowing water, with probably about 50% of Scottish trout production. From this it is possible to estimate that flowing water production in England and Wales amounts to between 11 000 and 12 000 tons, and throughout the UK the total is around 14 000 tons.

Water quality regulation is undertaken in England and Wales by the Water Authorities, and in Scotland by the Purification Boards. Although the structure of these organizations is different, there has been a trend toward a uniformity of approach which could be summarized as:

(a) The use of differential limits on river fed systems.
(b) The intention that the river quality as measured by using DoE or EEC quality criteria is not decreased.

Thus, as fish farms are normally based on class 1 rivers, the existence of fish farms should not reduce this to class 2, and where the river class is 1A, this in turn should be maintained. Although in some cases a reduction to class 1B for a short distance downstream may be permitted if recovery then takes place.

Fish farms are now regulated under the Control of Pollution Act 1974, and, where relevant, the Acts that preceded it. Under this legislation, authorities are required to grant discharge consents which normally include volumetric limitations and definition of specified parameters.

In the majority of Consents the parameters listed in Table 9.2 are set with the differential range shown.

Table 9.2. Consent parameters

	Differential-range (mg/l)
Suspended solids	4–10
B.O.D. (nitrification inhibited)	2–5
Ammonia	0.4–1
Dissolved oxygen saturation	50%–60% minimum absolute

For dissolved oxygen, absolute minima are normally set, although relief from these absolute minima are sometimes granted where dissolved oxygen of the in-coming water is highly variable and may be low on occasions.

The discharge consent normally includes volumetric criteria as it is the mass flow of the constituents of the discharge rather than merely their concentration that is important for the receiving stream.

Looked at in purely qualitative terms, fish farm effluents are normally of excellent quality and in themselves would often conform to Class 1, or even Class 1A, definitions. The volumes of effluent, however, are normally very large, and often represent a very significant percentage of river flows, especially at dry weather minima. Thus when mass flow calculations are made, fish farming on flowing water systems may add substantially to the loading of the river.

Additionally, in the setting of Discharge Consents a number of other items are often specified such as turbidity and the concentration of specific chemicals and heavy metals.

In England and Wales the privatization of the water industry has led to the establishment of the National Rivers Authority (NRA) which will take over responsibility for water quality and therefore presumably the monitoring of all Discharge Consents, including those of fish farmers. It is expected that the NRA will follow the general principles that have been established by the Water Authorities.

In the paper to EIFAC working party on Fish Farm Effluents presented on 9 May to 1 June 1987, Solbe reported that production of solids from the UK fish farms which had been measured at 1350 kg of dry solids per ton of fish produced in 1980 was calculated at

546 kg per ton in 1987. This represents a significant improvement, and, notwithstanding the reservations expressed, appears to be in line with figures produced by a number of other European countries. If this figure is applied to 14 000 tons of production in flowing water and it is assumed that a flow of 5 litres per second is taken for each ton produced annually, this would equate to an increase of 3 mg/l of suspended solids, assuming that all solids production was discharged immediately into the watercourse. Unfortunately, this may be too simplistic a solution as solids production is higher during the warmer months of the year when fish are more intensively fed, and this is also the time when river flows are at their minimum. However, the returning flow from the fish farm will be diluted by the by-pass flow remaining in the river, so this remains an important factor in setting of Discharge Consents.

A major reduction in solids production within fish farms must be credited to improved feed formulation and increased digestibility which has come about as a consequence of a better understanding of fish nutrition. There is every reason to believe that this trend has some potential to continue in the future, although more work will need to be done to establish alternative sources of digestible protein, other than fish meal.

The present regulatory regime and the parameters set provide a reasonable basis for the regulation of fish farming within the water environment, and indeed it would be difficult to justify a further tightening of Consent limits in most cases, especially as it must be recognized that the methods of sampling and measurement are, for many parameters, already approaching their limits both for accuracy and for discrimination. This does not, however, mean that even at these levels fish farming has no cause for concern nor other responsibilities to meet.

It is often the visual impact which is a cause for concern, and matters of general housekeeping have also given rise to difficulties with other riparian users. In a report prepared by Ian Allen for the Directors of the Test and Itchen Association on the impact of fish farming on those two rivers, a number of other points were brought up. These included the reduction in stream and river quality, where between the point of abstraction and return, the flow in the river itself was substantially reduced, the problems of providing free passage of migratory fish both up and down stream, the question of escapees from the farm into the wild, and the disposal of the inevitable mortalities that occur on the fish farm in a way that is acceptable and inoffensive to those enjoying the surrounding area.

To these concerns can be added the appearance of suspended solids leaving the fish farm even where they are well within Consent limits, and it will be necessary for the fish farming industry to respond to these concerns in a constructive manner.

The question of flow reduction in channels will depend on the volumetric component of any discharge consent, and must be a factor considered when this is granted. While it is not possible to lay down precise criteria applicable in every case, it cannot be acceptable to reduce the original river to a trickle or indeed dry it up in extreme cases. Matters will be made worse where the extraction and discharge points are a considerable distance apart, thus affecting a long length of river; the distances should be kept as short as possible.

The passage of migratory fish is bound up with residual flows in main river channels, and will be a factor to be considered. The downward migration of smolts normally presents a lesser problem, but it is important that fish farms are designed so as not to create smolt traps, and that screens are suitable to prevent the entry of smolts into the fish farm itself.

Escapees from fish farms have presented problems, especially on more important game fishing rivers. This problem can be overcome by the proper design and installation of screens, and has probably on balance become less serious as the standard of construction and management of fish farms has become more professional in recent years.

The problems of antibiotics used in fish farms have also been cited in recent years, although present research has not identified this to be a major threat, not least because the compounds used are not used in human medicine, and measurements made on residues indicate that these decrease very rapidly over quite a short distance downstream from the fish farm, even where a measurable concentration existed in the effluent.

Because fish farm effluents are already of high quality in absolute terms, methods available to improve them are limited, and the only method relevant is by primary settlement, although it is possible that more sophisticated methods of treatment could be justified in a fish farm which recycled its water and hence used it much more intensively. The results of the construction of settlement ponds for fish farms has been variable, as have been the results of some of the more sophisticated systems, such as swirl concentrators. There is a need for more information and research on treatment methods applicable to these effluents which already contain such low concentrations of solids, as there is a general awareness in the industry of the need to improve the quality of returning water, especially in visual terms, even where it complies with existing Consent conditions.

The NFU of England and Wales are revising and improving their Codes of Practice relating to fish farming and the use of chemicals in fish farms. It is the intention that these statements contain more quantitative data than hitherto, and will lead to much more precise advice being given.

Although the outputs of fish in a cage culture system are the same, the problems are somewhat different, and Discharge Consents are not applicable to non-flowing water conditions. Because of the very slow water exchange, changes are likely to be of a long-term nature, and they include the deposition of a layer of waste matter below the cages and the release of nutrients and trace elements into the systems which may result in significant eutrophication of the water in many lochs and upland lakes.

Models to describe the behaviour of these waters in the event of the introduction of a fish farm have been created.

There is no doubt that increased attention will be paid to these matters as the demand for sites for fresh-water cage culture intensifies, not only for trout production but to meet the increasing demands for smolt production for the Atlantic salmon farming industry.

CONCLUSION

Fish farming is a legitimate user of high quality water for food production. Its output has expanded very considerably in the last fifteen years to meet the demand of consumers for a healthy and nutritious product. Fish farmers are aware of the need to co-exist with other water users and to take a responsible attitude to the way in which they operate their farms.

The existing regulatory framework provides an adequate basis for this to continue, and there is no evidence of widespread or significant damage to the water environment from fish farming operations within the UK.

The scope for increasing production in fresh water within the UK is now limited by the availability of additional water and sites, and whilst likely to grow, is unlikely to show the rates of growth achieved to date. Marine fish farming has, however, considerable potential for further growth.

There is every reason to believe that the water needs of fish farming can be met without detriment to other users, and that the industry can co-exist amicably with them.

10

Pollution from fish farms

J. G. Jones, BSc, MSc, CChem, MChem A, FRSC, (Member) [†]

ABSTRACT

Fish farms are continuing to grow in numbers and in size. They cause concern because of their location in areas of high-quality water, frequently in the head-waters where there is little dilution for large volumes of effluent. This jeopardizes the water quality, and may affect the ecology of the river—migratory fish in particular.

The use of chemicals for the treatment of disease is causing concern, particularly if the river is used for potable abstraction. Little information is available on the low-level effects and the detection of chemicals such as antibiotics and hormones. The chemicals are not controlled nationally, the only control being through consents.

With the continued growth of the fish-farm industry, problems are likely to increase unless a responsible attitude to their development is adopted.

KEY WORDS   Development   Pollution   Abstraction   Dilution   Chemicals   Disease   Environment   Control   Legislation

INTRODUCTION

Fish-farming development can have an effect on the environment, and, in particular, on a watercourse. A recent joint publication [1] contains the following reference to fish farming:

> The Government has announced its intention of extending water abstraction licensing by water authorities under the Water Resources Act 1963 to fish farms rearing fish for the table. This will allow water authorities to exercise greater control over river flow and will supplement existing requirements for all inland fish farms to obtain discharge under the Control of Pollution Act 1974.

The report, in summarizing the water authorities' comments on various aspects of pollution from farm wastes, states that:

> Pollution arising from fish farming has not been considered in previous reports. The Wessex, Welsh, and Yorkshire Water Authorities have reported serious problems associated with fish-farming activities. They have pointed out that,

[†] Head of Scientific Services, Wessex Water.

although the number of reported pollution incidents is very small, they may not reflect the considerable damage improperly managed intensive units can cause to other water users. This damage can be exacerbated through over-abstraction of river water, leaving the residual flow in the river too low to adequately protect aquatic life or provide dilution for the resulting effluent containing fish faeces, excess fish food, and sometimes potentially harmful chemical formulations.

The water authorities consider that this unsatisfactory situation at some farms has stemmed from the present inadequate legislative control. Fish farms producing food for the table are exempt from normal planning controls and abstraction licensing, even if this means completely drying up short stretches of river systems. The water authorities have therefore welcomed the announcement from the Government of its intention to extend water abstraction licensing for such fish farms.

FISH FARMING IN WESSEX REGION

The commercial farming of rainbow trout comprises two basic processes: hatching, and rearing of young fry to 'table' size. Most trout rearing is carried out either in earth-dug ponds or fibre glass tanks. Water supply may be derived from rivers, springs, or groundwater, and feeding is usually based on high-protein dried-food diets.

Although in 1972 the UK had the sixth largest trout-farming industry in Western Europe [2], this represented only 6% of the total Western European trout production; France, Italy, and Denmark together produced 65%. Between 1976 and 1981, however, the 'table' trout industry in the UK increased by 330%, with most of the increased production in England taking place in waters derived from the chalk ridge which runs from Yorkshire to Wessex [3]. This expansion has mainly taken place in the areas of four water authorities: Yorkshire, Thames, Southern, and Wessex.

There are 55 fish farms in the Wessex region, of which 17 are in the catchment of one major river [4]. Large fish farms can require over 150 Ml/day of high-quality water which becomes contaminated by the waste products of fish and excess fish food. The treatment of such large volumes of dilute waste, produced by a typical fish farm, can give rise to potential on-site treatment problems. In addition, the abstraction of a large proportion of river flow not only removes the available dilution in the main river but leads to obstruction in the movement of migratory fish and the entrapment of salmon smolts and coarse fish fry in the fish farm.

The Southern Division of Wessex Water is particularly suitable for the production of rainbow trout. It contains clear, well-oxygenated rivers which are often derived from chalk springs. The temperature conditions in the south of England and the alkaline river waters are believed by fish farmers to be important for the rapid rearing of trout. Also, the close proximity to markets makes rivers such as the Hampshire Avon attractive for trout farming.

The Avon catchment has seen a dramatic increase in the number of trout farms which it supports, and probably now produces more trout than any other English river catchment. More large farms are planned for this river, which in its middle reaches, between Salisbury and Christchurch, supports three large fish farms whose combined rainbow trout production exceeds 1000 tonnes per annum.

Following growing concern about the pollution of rivers caused by fish farms, in 1982 the National Water Council and National Farmers Union set up a working party with the main objectives of improving the operation of fish farms and attaining a common approach to setting consent conditions for fish farms. The working party successfully completed its work toward the end of 1983, and the agreed findings resulted in a Code of Practice. The recommendations go a long way to ensuring that existing fish farms are properly managed so that many of the previous problems, such as desludging of ponds into the river, do not occur. However, there were a number of major points which such a working party was in no position to determine. These include control over abstraction and the safety of chemicals which are available for use by fish farmers.

POLLUTION FROM FISH-FARMING ACTIVITIES

The major pollutants released from fish farms are derived from unconsumed food and faecal wastes from the fish. The effluent from fish farms contains organic material which exerts a biochemical oxygen demand (BOD) on the river water. In addition, it will contain ammonia and suspended solids. These are the constituents which have to be controlled in any discharge of sewage effluent. It may be useful, therefore, to compare the pollution load of effluents from fish farms with those from sewage-treatment works.

In France it has been calculated that one tonne of trout on a fish farm can produce a pollution load equivalent to the treated sewage effluent produced by 200–500 people [5]. In 1982, from a comprehensive survey of UK fish farms, staff of the Water Research Centre calculated that the BOD for each tonne of trout produced per annum was equivalent to that in the effluent from a sewage-treatment works serving 314 people (or untreated sewage from 20 people). They also showed that for each tonne of trout produced per annum, the ammonia load was equivalent to that from a sewage works serving 122 people, and the suspended solids load was equivalent to the output from 859 persons. It was noted that the discharge from a fish farm in terms of BOD or suspended solids could be equal to the output of treated wastes from 'a sizeable town'. Using these figures, a production of 1000 tonnes of trout each year would give a BOD load equivalent to that produced by the treated effluent from a town of over 300 000 people or the untreated sewage from 20 000 inhabitants [6].

Fish diseases

There is uncertainty and concern about whether fish-disease outbreaks at fish farms can affect wild fish stocks in rivers. A study group which has examined this problem concluded that wild trout populations could become infected with the infectious pancreatic necrosis (IPN) virus escaping from an infected fish farm. It is possible that the combination of stress (caused by fish-farm pollution) and the exposure to high concentrations of disease agents (released from infected fish farms) could lead to disease outbreaks in wild stocks. More evidence is urgently needed.

Chemicals used in fish farming

The use of chemicals on fish farms is another area where concern has been expressed, particularly with respect to the use of antibiotics and antibacterial agents. However, the methods of analysis for measuring these substances are not readily available. Three potential problem areas have been identified, namely (a) their mutagenicity, (b) their ability to produce allergic responses, and (c) the possibility of the development of resistant strains of micro-organisms.

Uses of Hampshire Avon

In addition to fish farming, the River Avon is extensively used as a source of water for public water supply. The possible effect of traces of these chemicals in water supply raises the question as to what concentrations of these substances are acceptable, and how they are to be determined. Therefore, more investigative work on this subject is required. At present the Ministry of Agriculture, Fisheries and Food is conducting work in this area, and their results are awaited with interest.

The Department of the Environment/Department of Health and Social Security Joint Committee on Medical Aspects of Water Quality has been consulted about the use of antibiotics and other chemicals at fish farms. They have expressed an interest in investigating the problem in greater detail, and being provided with further information.

In the Wessex Water region, owners of fisheries downstream from fish farms have often complained about the escape of small rainbow trout and their effect on the balance of fish stocks, and thus the quality of their fishing. In particular, on the River Wylye (a tributary of the Avon), escaped rainbow trout from an upstream fish farm have consistently evoked complaints from the downstream fishery owners who claim that the natural brown trout fishery has been badly affected [7].

A recent independent study, carried out by the Freshwater Biological Association on the Hampshire Avon to assess the fish populations downstream from a major fish farm, was unable to support the view that large number of escapers of rainbow trout were present. In fact, not a single escaper was caught.

Where a high proportion of the flow in a large river is diverting into a fish farm, the draw of water is likely to lead downstream migrating salmon smolts into the farm ponds where they could suffer considerable mortalities. It is widely agreed that it is difficult to efficiently 'fence' or block rivers to movement of fish, and consequently this possible loss of young salmon is now giving cause for concern. At present no easy or inexpensive answers to the problem appear to be available.

Work carried out by the Wessex Water Fisheries staff, in conjunction with staff at the Trafalgar Fisheries at Longford, has assisted in developing techniques to reduce the numbers of fry which have become trapped in the fish farm.

ENVIRONMENTAL IMPACT OF FISH FARMS

In managing the water quality of rivers, Wessex Water has used the quality criteria of the National Water Council (NWC) river classification to set long-term water quality objectives. In particular, in the long term almost all the Hampshire Avon and its tributaries are to be maintained as a Class 1A river. Evidence is unfolding that below some fish farms this classification is not being met.

A Water Research Centre (WRc) report [6] noted that data from a range of fish farms showed an average increase in river water BOD through a fish farm of 1.5 mg/l. This has also been confirmed in examination of data from farms in the Wessex Water region. Also, an examination of the quarterly average BOD concentrations in rivers above and below fish farms frequently shows a significant increase in BOD below the fish farms. It is this type of water-quality impact which gives cause for concern, particularly where it is necessary to maintain a Class 1A river.

The WRc report [6] found that, on average, UK fish farms increased the concentration of amm N downstream by 0.17 mg/l. The effect is evident when fish farms in the Wessex region are examined. River-water samples taken downstream from a fish farm on the River Wylye consistently show increased levels of amm.N which frequently exceed the NWC Class 1A 95 percentile limit of 0.4 mg/l.

Below fish farms in Wessex Water's region, there is normally a reduction in the dissolved oxygen (DO) concentration of the river, caused by the respiration of the fish and the oxygen demands of their waste products. Solbé's survey of UK fish farms [6] recorded an average decrease in DO of 1.6 mg/l. In the case of fish farms in Wessex, this figure is commonly exceeded.

River-water quality deterioration below fish farms at present affects only a limited length of river with respect to chemical and biological criteria. However, the future developments which are being discussed could be adversely affected unless a major change occurs in the treatment processes which take place on fish farms.

A recent survey of four major fish farms in the Hampshire Avon catchment showed that organic solids and sometimes sewage fungus were present immediately below the fish farms, and in all cases the macro-invertebrate fauna of the river bed was affected because it contained more pollution-tolerant organisms. The following statements list some of the effects which have been noted in recent surveys:

Fish farm A
A great depth of silt deposited in the lower reaches of effluent channels. Biological examination showed a small effect on river life which extended about 500 m downstream.

Fish farm B on Hampshire Avon
Tributaries of the Hampshire Avon, into which fish farm effluent is discharged, show a significant effect on the ecology—with the community of animals in these channels being dominated by Asellus and chironomids. The channels contain a thick layer of greyish organic material with a prolific growth of sewage fungus in one channel, and there is an observable effect on river life immediately below the point where the effluent channels rejoin the main river. The river returns to normal 1 km downstream.

Fish farm C on Hampshire Avon
An effect on the aquatic life of the river was evident for about 0.5 km downstream.

In assessing the impact of fish farms on the aquatic life of a river, frequent use is made of a 'biotic score' system which is designed so that the pollution tolerance of the animals in the river life community can be described numerically. For example, leeches and midge larvae score low, whilst mayflies and stoneflies score high. Using the system which was devised for the 1980 national NWC River Quality Survey (known as the

BMWP score system), a study of the entire River Avon was carried out. Fish farms, sewage-treatment works, and a few industrial discharges, all discharging to various parts of the river, are controlled by existing legislation. However, the only foreseeable major increase in effluent discharges will be that from fish farms. With the proposed future development in fish farming, even with stringent consent conditions, it is probable that discharges will result in a deterioration of the biological and chemical quality of the Hampshire Avon over an increasing stretch. Therefore more stringent controls are needed.

RIVER FLOWS AND FISH FARMING

Fish farms producing fish for the table (which includes most fish farms) are classified as agricultural units and, are normally exempt from licence control on river abstractions. Because fish farms use large quantities of water, without effective control of volumes abstracted, there is the potential to cause serious problems in rivers.

A fish farmer could abstract the whole of the river flow between his inlet and outlet, leaving a river with virtually no flow, for example on rivers such as the Hampshire Avon where there are many channels and millstreams controlled by hatches. Flows can be diverted down one channel to supply a fish farm—substantially reducing flows in parallel channels. These changes can seriously disrupt aquatic life in the river and, where most of the flow passes through a fish farm, it could present a barrier to migratory movements of salmonid and coarse fish. The lengths of the river over which flows are affected are likely to become unsuitable spawning areas for migratory fish, as there would not be the river flow to keep the spawning beds clean.

Already, several farms abstract most of the dry-weather flow of the river, and, with continued expansion of existing units and the establishment of new farms, the situation will become more difficult [8, 9]. One farm on the Hampshire Avon abstracts about 70% of the average summer dry-weather flow, which is more than the minimum recorded flow. As a result, there are about 2 km of the river channel which can have a severely depleted river flow during the summer and autumn.

At one point, the river is divided between three main channels, with the flows to each controlled by hatches. There were proposals to establish a fish farm with an initial abstraction of 90 Ml/d, from the main channel just below one of the set of hatches, with control of the hatches being exercised by the fish farmer. The 95 percentile low flow of the Avon at this point is 325 Ml/d, with a minimum recorded flow of 154 Ml/d. In order for the fish farmer to operate during the summer, he would have had to divert most of the river flow toward his fish farm. This would deprive the flow in the eastern branch of the Avon, leaving inadequate dilution for the effluent from a sewage-treatment works. The quality of the river downstream would deteriorate, and there is little that the Authority could have done to prevent this happening. Fortunately, the proposed development was rejected for an alternative site where the same problems exist but to a smaller degree. A new fish farm was opened in late 1983, but the owner is proposing further expansion, therefore problems may yet arise both from the conflict on river flows and the additional organic load on the river.

When the Water Resources Act was implemented in 1963, all agricultural abstractions from rivers were exempted from abstraction licence controls, except spray irrigation which was put under control because of the potential impact on rivers. The fish-farming industry was almost non-existent at the time and was not considered in the Act. During the last twenty-five years there has been a major change, and fish farming is now an important and rapidly growing industry. There are good reasons for such developments to be covered by licences, which would help to avoid conflicts with other river users and prevent serious changes to river ecology. Therefore it is important to the environment that the Government's proposals to extend abstraction licences to include fish farms are carried to fruition.

CONCLUSIONS

(1) Evidence indicates that, where the volume of discharge is a high proportion of the flow downstream, fish farms have a significant effect on the chemical and biological quality of the receiving watercourse for a limited distance downstream.

(2) This situation might be accepted for a few fish farms which are well separated along a river. On some rivers fish farms are being developed in close proximity to one another, and this may affect the quality of water over significant stretches of river.

(3) The existing legislation is satisfactory for the control of water pollution from normal discharges. Fish farm effluents can be difficult to control because of the small dilution with clean water and because they are often discharged to high-quality rivers where a relatively small deterioration will result in downgrading.

(4) The absence of control of the amount of water abstracted by a fish farmer will result in lengths of the best-quality rivers having significant changes to their river flows, and this must have a deleterious effect on the ecology of these rivers. This condition is likely to adversely affect migratory fish and, one hopes will soon be rectified.

(5) Some of the chemicals which are used by fish farmers could cause concern to downstream water users, and, in particular, abstractors for public water supply. The control exerted at present on the use of these chemicals does not appear to take account of downstream water use. A voluntary agreement scheme such as the Pesticides Safety Precautions Scheme, where the properties of new chemicals are investigated before their use is agreed, could usefully be extended to include chemicals used by fish farmers

ACKNOWLEDGEMENTS

This paper has been prepared from various reports, papers, and surveys carried out by staff, past and present, of Wessex Water. The views expressed here do not necessarily reflect the views of the Authority.

REFERENCES

[1] Water Authorities and Ministry of Agriculture, Fisheries and Food (1987) *Water Pollution from Farm Waste 1987, England and Wales.* 1987.7.

[2] Lewis, M. R. (1980) Rainbow trout: production and marketing. Misc. study No. 68. Department of Agriculture Marketing and Economics, University of Reading.

[3]   Needham, T. (1982) Trout farming in the UK—the way ahead. *Symposium on Commercial Trout Farming.* Institute of Fisheries Management, University of Reading.

[4]   Wessex Water (1984) An appraisal of the effect of fish farming on river quality in Wessex.

[5]   Faure, A. (1979) Mise au point sur la pollution engendrée par les pisciatures. *Pisciculture*, **13**, 33–35.

[6]   Solbé, J. F. (1982) The nature and effects of fish farming effluents. *Symposium on Commercial Trout Farming.* Institute of Fisheries Management, Reading University.

[7]   Allan, I. R. H. Report to the Test and Itchen Fishing Association on a study of the impacts of fish farming on the fisheries and fishing in the Rivers Test and Itchen, Hampshire. *Consultant's Report.*

[8]   Davies, R. (1979) Anti-pollution laws could wipe out small producers. *Fish Farmer*, Jan., 19–20.

[9]   Warrer-Hansen, I. (1978) Fish farms are having to watch their waste. *Fish Farmer*, Sept., 19–20.

11

Water source protection and protection zones

M. B. M. Harryman, DPhil, BA [†]

ABSTRACT

This paper discusses Government policy to August 1989 on the use of protection zones to protect UK water sources from pollution by nitrate. It outlines the use of the powers in the Water Act 1989 to declare Nitrate Sensitive Areas, and the procedures that would be followed in declaring such zones. The circumstances are outlined in which compensation would be paid in Nitrate Sensitive Areas and the reasons for departing from the 'polluter pays' principle. UK policy is related to the proposed European Community Directive on the control of nitrate, and the similarities between the two approaches are illustrated.

KEY WORDS   Nitrate sensitive areas   Water protection zones   Diffuse pollution   Water Act 1989

INTRODUCTION

This paper considers the use of Water Protection Zones as one element of the Government's pollution control strategy.

The quality of surface water sources in this country is generally high, reflecting the success of measures over a number of years to control point source discharges. Protection zones provide a mechanism that can be used to limit pollution from diffuse sources. New strengthened and simplified powers to declare protection zones were provided by provisions in the Water Act which received Royal Assent in July 1989. This text has taken account of legislative changes between March 1989 and August 1989.

In this paper, consideration is given to: (a) control on nitrate as a practical, although not necessarily typical, example of the way in which protection zones can be used; (b) the background to the current UK policy on the control of nitrate; (c) the factors influencing the development of that policy and (d) recent developments that will affect the use of protection zones to control nitrate in the UK. The paper does not consider other potential use of protection zones such as, for example, protecting vulnerable water supplies from other pollutants (for example pesticides) although, of course, they can be used in such cases. In the Water Act, protection zones for the control of nitrate are referred to as

[†] Water Environment A. Division, Department of the Environment.

Nitrate Sensitive Areas (NSA), and that formulation will be used throughout the paper.

THE PROBLEM OF DRINKING WATER SOURCES

Drinking water supplies are required to meet the standards set in the Water Supplies (Water Quality) Regulations 1989, which include those set by the European Community Directive concerning the supply of water for human consumption. This Directive provides, *inter alia*, that the level of nitrate in drinking water must not exceed 50 mg/l. At the beginning of 1988, the Government announced that existing derogations allowing water undertakers to continue supplying water above this limit were being withdrawn. This decision did not reflect any change in medical judgement of health risks. It was taken because of legal advice indicating that the Directive did not allow derogations to be given in the particular circumstances which were causing the nitrate limit to be exceeded.

Most supplies already comply with the limits in the Regulations and the EC Directive. Where they do not, authorities have programmes based on blending and treatment to ensure that they will comply within a reasonable period. But these measures do not by themselves prevent nitrate levels rising in areas where there is intensive arable or mixed farming. It is these areas where the use of NSAs can be expected to have a role, and are being considered as an integral part of the strategy for longer term protection of sources.

PROBLEM OF NITRATE IN WATER

There has been extensive research into the levels and sources of nitrate in water, and into the action that could be taken to control nitrate levels. In 1986, for example, the Department of the Environment (DoE) published the conclusions of the Nitrate Co-ordination Group in a report 'Nitrate in water'. The Group's membership included representatives of Government, the water industry, agriculture, fertilizer manufacturers, and scientific research organizations. The group reviewed the state of knowledge on nitrate, and made recommendations for protection policies and for further research. They considered the role of protection zones, the options for the water industry, and the scope for restrictions on agriculture.

More recently, in 1988, DoE published published in 'The nitrate issue' the results of a further study. This considered the economic and other consequences of various local options for limiting nitrate concentrations in drinking water. Ten areas were selected as suitable for a theoretical study to provide a good spread of differing circumstances. The effectiveness and cost of the various options for action were assessed for each area. The study concluded that each area had special characteristics which would need to be taken into account before any decision can be reached on the most desirable and practical options. In some cases preventative options would not be effective until the second half of the next century. Often the cheapest option locally was not the cheapest overall for the country.

In November 1988 the Minister of Agriculture made a statement about the Government policy to control nitrate. The Minister drew on the results of these studies and mentioned the complexity of the nitrate problem. He announced that the Government expected that remedies might vary considerably from area to area depending on local circumstances such as geology, rainfall, and farming patterns. In some cases, water blending or treatment might provide the best option for reducing nitrate levels in drinking water supplies—and indeed the only one in the short term. In other cases, agricultural restrictions might be preferable, or a mixture of water and agricultural measures. The Minister went on to announce that the Government believes that, wherever possible, any agricultural restrictions should in the first event be on a voluntary basis. The Government, however, considers that it is necessary to retain compulsory powers to establish as a fallback protection zones within which activities can be regulated. The Government also believes that, where farmers are obliged to restrict their agricultural activities beyond the degree which could be regarded as good agricultural practice, they should be compensated. An announcement about these compensation arrangements was made during the passage of the Water Act. A consultation paper 'Nitrate sensitive areas' was issued in May 1989 and views sought on the arrangement for compensation from water, farming, and other affected interests.

For most pollution policy in this country, the 'polluter pays' principle applies to any control regime, both for cleaning up damage to the environment, and for measures that satisfy environmental objectives. In the case of nitrate the Government recognizes that the situation is exceptional, and there are special circumstances that do not and cannot apply to other forms of water pollution. Firstly, nitrate is a natural and necessary prerequisite for agriculture, and without it plants cannot grow. Secondly, it has been the policy over many years to encourage agricultural production and with it the use of fertilizers. Thirdly, nitrate leaching is a function of the agricultural process of which fertilizers application is but a part; it may arise from activity some considerable time ago. The chalk aquifer in Southern and Eastern England, for example, has a response time of up to 40 years. It is these exceptional circumstances that have led the Government to conclude that some agricultural measures to control nitrate levels in water should attract compensation.

PROVISIONS OF THE WATER ACT

The 1986 consultation document 'The water environment—the next steps' set out the Government's view that it would be wise to amend Section 31 (5) of the Control of Pollution Act to facilitate designation of protection zones. The existing provisions were widely thought to be inoperable. The provisions in the Water Act provide simplified and more flexible powers.

Under the Water Act provisions the National Rivers Authority (NRA) will be responsible for identifying an area within which a NSA may be required. The NRA generally will be responsible for safeguarding surface and underground waters within a framework of statutory quality objectives. The Government believes therefore that the NRA will be best placed to identify the areas in which nitrate control measures will be needed. Proposals for a NRA will be published, and there is provision for objections to be lodged. Government may, if it considers it appropriate to do so, hold a Local Inquiry before making any Order on the application for a zone. It is likely that the Government will wish to hold Local Inquiries for at least the first zone, and for any subsequent proposals for designation where any new issues of principle arise that have not been considered before in public.

These powers strike a balance between the needs of water quality objectives to declare a NSA and the rights of the individuals affected by the proposals. The key point that must be considered in any discussion of NSA policy is that there are these two sides, and both must be treated with equity. On the one hand there is the need for safeguarding water supplies. On the other hand is the need to provide those affected by the controls with a means of representing their interests and influencing the policy to be adopted.

The types of restrictions in NSAs could introduce significant limitations on the choices available to farmers in the affected areas. Although compensation will be paid for significant measures that go beyond good agricultural practice, severe restrictions may be introduced, and those affected must therefore be allowed an opportunity to represent their interests.

Bringing this policy into effect opens up a number of questions to which answers will be needed over the next year. For example: the locations of the first zones; their sizes; how best to evaluate the different options available; and how long would be needed to assess the effects of voluntary measures?

EUROPEAN COMMUNITY ACTION

The European Commission has recently published a proposal for a Directive concerning the protection of fresh, coastal, and marine waters against pollution by nitrate from diffuse sources. The proposal would require Member States to designate areas as 'vulnerable zones' where the land drains directly or indirectly into:

(a)   Surface freshwater intended for abstraction of drinking water which could contain more than 50 mg/l nitrate if protective action is not taken;
(b)   Groundwater intended for abstraction of drinking water which contains more than 50 mg/l nitrate or could contain more than 50 mg/l nitrate if protective action is not taken;
(c)   Natural freshwater lakes, other natural freshwater bodies, estuaries, coastal waters and seas which are eutrophic owing to nitrate or which in a short time may become eutrophic if preventive action is not taken.

Within vulnerable zones, Member States would be required to introduce controls in regulations on the application to land of chemical fertilizer at levels to be set by national governments, and livestock manure at levels set in the Directive; and to establish rules governing the application of manure, and the design and capacity of manure storage facilities. Member States would be required to ensure that sewage plants serving more than 5000 people discharging into vulnerable zones designated to prevent eutrophication limit effluent to 10 mg/l nitrogen. Member States would also be required to consider further agricultural measures to reduce pollution by nitrate from diffuse sources.

The Commission regard the nitrate proposal as complementary to the Directive on the supply of water for human consumption. The new Directive would provide protection in the longer term for water sources. However, the European Commission sees these proposals as the first step to a comprehensive set of controls over nutrient inputs to surface and ground water and further proposals are expected to be published soon covering treatment standards for waste discharges.

The proposals for the nitrate Directive were published at the beginning of January 1989. They have received only a preliminary consideration by the Member States, and negotiations have only just begun. The UK has welcomed the Commission's initiative in bringing forward these proposals. However, the UK has reservations about some aspects of them. For example, we have particular concerns about the way in which the 50 mg/l limit is expressed in the proposals. In the UK about 68% of drinking water comes from surface sources. This is the highest level in the EC, outside Eire, and in some parts of the country, notably the Southern Water region, much higher proportions of ground water are used. The natural variation in the level of nitrate in surface water sources gives rise to special problems in meeting the terms of the proposals. We shall be looking carefully at this aspect of the proposals.

We would also wish the Directive to take account of the experience in the UK on controlling nitrate levels and the policies already in place. A significant difference between the UK and European policy which will need to be considered is that the current policy in this country is to adopt voluntary rather than compulsory measures to bring about reductions in nitrate levels.

CONCLUSIONS

There are similarities of approach between the policy for the control of nitrate being proposed by the European Community and that being developed in the UK. Community policy will rely on protection at source through restrictions. We are now introducing similar measures in the UK in the form of nitrate sensitive areas. However, the Government recognizes that blending and treatment will often be required, at least in the short term, to ensure drinking water of the appropriate standard.

In general, the Government favours a pragmatic approach to control of nitrate. The UK policy places more emphasis than that proposed by the European Commission on adopting solutions that are appropriate in different areas. Further details of the UK policy on the control of nitrate will be announced soon, and the Government expects to introduce measures in this country in advance of the EC Directive coming into effect.

Discussion on Papers 8, 9, 10, and 11

PAPER 8

Mr C. J. A. Binnie (W. S. Atkins and Partners), opening the discussion, said that the author had set out the usage of water based on the Penman equation. He said that there were several other similar computer programs that had been developed including that by Silsoe College. He asked if the authors' program differed in any material way from these?

He said that the licensed abstraction for spray irrigation far exceeded the requirement in normal years. He suspected that there were many farmers in the irrigable area who, through lack of access to water or lack of irrigation facilities, did not irrigate at all. He asked if anyone knew what the normal year usage of irrigation was?

Mr Binnie said that irrigation areas were basically in the drier south and east of the UK. In many of these there was competition for the water resources—water supply, dilution for effluent, fish farming, and the ecology. He said that he was concerned that there was conflict between irrigation abstraction and other users' requirements.

He felt that the new NRA must ensure a proper balance between the use of water resources. He asked NRA staff present to comment on this aspect and whether the situation would be reviewed?

Mr Binnie said that at present it was proposed that the NRA would be funded by Treasury and from the charges for abstraction and discharge. That meant that at present there was a very small charge for an irrigation abstraction, and the amounts taken were difficult to monitor. He asked the NRA if consideration had been given to charging substantial amounts, like the marginal cost of water in the river, for the amounts that were actually in the abstraction licences? If this were the case then the farmers might very quickly request a change of licence, thus effectively releasing water for other uses.

Mr N. C. Oxley (Howard Humphreys) understood that Dr Bailey had calculated the irrigation requirement solely on avoiding any stresses in the plant, and he wondered whether or not in the UK one would allow an additional amount for leaching to prevent any build-up of salts in the soil?

Secondly, he referred to the use for planning purposes of the fifth driest year in twenty. He said that he would rather see something presented in terms of probability, for example overseas one would conventionally use an 80% probability in terms of irrigation application. He asked for Dr Bailey's comments on what the fifth driest year represented in terms of probability?

He said that one should not relate the actual use to the licensed abstraction. He quoted figures for abstraction licences for East Africa where, in one specific river, the total volume licenced was about 35 times the mean flow in the river. There were individual licences which were in excess of the DWF in the river, and actual abstraction was consequently far below the licensed volumes.

Mr T. R. E. Thompson (Soil Survey and Land Research Centre) said that as he understood it the calculations for irrigation demand were based on an assumption that crops rooted to their optimum depth. He said there were certainly cases where that was not true, and in his experience those were not uncommon. He asked to what extent that would alter the ultimate calculation of irrigation demand?

Mr R. W. D. Franklin (Yorkshire Water—NRA Unit) asked about the way in which the author had classified the country into seven areas. He said that for the Yorkshire area a rainfall station in Shropshire had been used. He wondered about the validity of this?

He said that the conclusions that Dr Bailey had reached did, however, bear out experience in Yorkshire. Typically, irrigation usage had been less than 50% of the total licensed quantities until 1984, which was the most recent dry year, when about 65% of the total licensed quantity had been used.

Mr Franklin said that in Yorkshire a system of guidelines had been adopted to control irrigation. This worked on a quota system. Also, time limits were now being put on irrigation licences—usually 10 years and 5 years in sensitive areas.

With regard to the NRA role he said that it must be remembered that the NRA nationally was functioning only in an advisory capacity at this stage, but the regional units were beginning to think about a co-ordinated licensing policy. Charging schemes for abstraction would have to be reviewed within the next couple of years.

Dr H. F. Cook (Wye College) said that he presumed that, in the MORECS approach, account was taken for interception loss from the canopy, and asked Dr Bailey to comment.

Mr J. Adlam (Travers Morgan) asked how often it was necessary to incorporate drainage systems with irrigation in the UK and overseas?

Mr I. Svoboda (West of Scotland Agricultural College) asked if the author could describe the restrictions on irrigation water quality? For example, would it be possible to use water from treated farm effluents?

Mr J. Morris (Silsoe College) said that Dr Bailey's model was largely physically based, and he asked if economic aspects of irrigation had been introduced into the equations, particularly relating to crop response and the costs of irrigation installation, and how they might affect the estimates of demand?

He asked if this kind of approach was useful from a management and policy viewpoint in view of predicted changes in the organization of the water industry. He asked Dr Bailey if he could see the water industry using this approach to plan and design catchment programmes possibly justifying investments in water supply for irrigation and possibly as a basis for water pricing?

Mr D. A. Burroughes (British Water & Effluent Association) said that reference was made in the paper to licences for over 200 million m^3, which was only likely to be approached in an exceptional drought year. Were there any provisions to call in unused capacity to allow new applicants some chance of obtaining a licence, for example if a cereal farmer switches to market gardening, fruit, or nursery gardening?

PAPERS 9 AND 10

Mr C. J. A. Binnie (W. S. Atkins & Partners), opening the discussion, said that he considered that tremendous strides had been made by the fish farming industry and he felt that they should be congratulated on what they had achieved. He said that he was surprised, however, to hear that they did not need a water abstraction licence and he wished to ask the NRA delegates present what their views were on this aspect?

He said that there was the possibility of significant effect from fish farm abstractions and discharges on the ecology of the river. Some effects were insufficient water remaining in the bypass channel; discharges to a separate section of river downstream; and increased BOD, amm.N and SS, and reduction of DO. He was concerned that, for example, where the BOD differential was between 2 and 5 mg/l per farm and there were, say, ten farms down a river, then a large increase in BOD could result—that equated to a sewage treatment effluent for a population of 100 000, and he would very much like to hear Mr Jones's comment in the presentation of the subsequent paper.

He believed that there were a number of treatment methods for treating effluents of that nature. He said that if the fish farming industry was serious about improving its effluents then he was sure that alternatives were available and that methods could be adopted to provide it with what was needed. He felt that it was up to the fish farming industry to initiate some discussion on this subject.

In conclusion he said that he found this paper, and the previous paper on irrigation abstraction, to be very interesting because they highlighted two competing demands for water. There were also of course requirements for water supply and for the river ecology. He believed that the public would expect the rivers to be looked after to a better extent than in the past, and he looked forward to hearing more from the NRA delegates present.

Mr H. Smith (Clyde River Purification Board) said that he was interested to read in the paper that the author considered that discharge consents were not applicable to non-flowing water conditions. He said that it might be interesting to bring into context the Scottish position. There had been a legal problem with the caged fish farm installations. The Clyde Board felt that they should be controlled but had serious doubts as to whether the waste being discharged could be defined as a trade effluent under COPA. The Highland RPB did not have such doubts, and started issuing consents. The Clyde Board sought legal opinion on the matter, and after lengthy deliberations it was decided that the wastes from such installations were indeed effluent and subject to consent under s.34 of COPA. Colleagues at the Western Isles, who were their own river purification authority, also sought legal opinion, and their QC said that it was not trade effluent and not subject to control. He realized that there were provisions in the Water Bill that should resolve this problem.

He asked Mr Saunders-Davies why he felt that discharge consents were not applicable to non-flowing water conditions?

In a subsequent contribution following the response from Mr Saunders-Davies, Mr Smith said that he was well aware of the NFU concern in this, and there was concern that one river purification board had imposed a condition on fresh water lochs, for example, that the phosphate level should not exceed 7.5 µg/l. The fish farm view had been to question how it was known that the phosphate was derived from fish farm discharge and that developments such as afforestation would not cause the phosphate level in the loch to exceed that. That had all to be taken into account.

He felt that his Board was adopting a sensible approach. It felt that there was need to control these discharges. The consents that had been issued related to the quantity of fish held in the cages. A requirement had been placed in the consent for the fish farmer to carry out monitoring and to make returns to the Board so that any changes could be detected in water quality. The Board would simply carry out an audit from time to time to ensure that such reports were accurate. It had been a very cautious approach.

Mr R. G. Toms (UK Centre for Economic and Environmental Development), opening the discussion on Paper 10, said that fish farms were very much in keepii.,) with agriculture—they fitted into the environment and were not unpleasant. He said that they had an excellent product which had added to the range of the consumer. On the other hand, he said that it was the intensification of fish farms which had led into problems. He felt that it was a similar story to farming as a whole, in that there had been encouragement to intensify, to achieve more product at lower cost. He said that intensification had taken place with little apparent consideration for its effect on the environment. He felt that much of this must come back to the MAFF because it had encouraged fish farming and encouraged improvements in yields of fish, but had provided no help in dealing with the environmental problems that that had produced.

Mr Toms said he felt that the present legislation was not suitable for the control of fish farms. A large fish farm could take up most of the river flow and consequently a very small increase of BOD or SS would reduce the river quality. Such a small increase could be within the experimental error of these tests. He felt that on the question of fish farms one ought to be looking much more at trying to control the effect on the river, and in doing that, more attention should be paid to biological indices rather than the chemical tests. The work done in the Wessex WA area had shown that there was a change in the biology of the river below a large fish farm.

He said that it became important when one appreciated that the Hampshire Avon was being used downstream of major fish farms for public water supply. He said that it was difficult to find out what chemicals a fish farmer might be using to improve fish stock. How could one protect the water intakes downstream if it was unknown what was being added to the river upstream? He asked Mr Crewe if there were any methods by which one could get such information? Mr Toms also asked the author if he accepted that there could be biological indicators which would assist in determining whether some of the chemicals being used, which might be biologically active, could affect water intakes.

Mr Toms referred to the treatment of fish farm effluent and said that with the small fish farm there was no particular problem—the environment could absorb most of them if they were well run. He said that it was the large ones which needed a greater degree of treatment. He asked Mr Crewe his views on the treatment of fish farm effluents.

Mr R. M. Walls (West Hampshire Water Company) said that it was his experience that there was not a great problem with the treatment of water taken from supply intakes downstream of fish farm effluent discharges. However, he said that what did concern him was the aspect of the biological chemicals, antibiotics and pesticides and, in some cases, steroids, etc. He said that, by and large, they did not appear in the EC Directive, apart from pesticides, but he felt that sooner or later it would have to be justified to consumers that they were not in the water, which would entail analytical input. To assist in that, he asked if fish farmers would divulge information on the types and quantities of chemicals being used and when?

Mr K. Wade (Welsh Water Authority) said that he wished to add his support to the comments made about the use of biological indicators to assess subtle changes below fish farms and environmental impact. He said that within Welsh Water a number of indicator systems had been produced specifically tailored for catchments and specific problems within catchments. There were now much more rigorous methods of producing objective indicator systems, and it was quite feasible that such a system could be specifically developed for measuring the impact of fish farm effluents on the Hampshire Avon. It might then be possible to write such an objective system into a legal consent to discharge.

In response to a query from the Chairman, Mr Wade confirmed that such systems were quick enough to be effective in monitoring and control. From extensive initial surveys to set up an indicator system, the indicators themselves perhaps might be 6 to 8 invertebrate taxa which could be identified quite rapidly. It would allow monitoring at quite a high frequency and be a cost-effective way of assessing the impact from a fish farm.

Mr J. Seager (Water Research Centre) said that one of the significant findings of the WRc study that had been referred to in the UK survey of fish farms was that there was a significant reduction in the amount of SS produced in 1986, compared to 1980, on the twenty or so farms from which there were reliable data available. One of the reasons that had been cited for this was that there was a significant improvement in the dietary regime. With that in mind, he asked Mr Crewe if in his area the Wessex WA was doing anything to encourage the fish farmers to use more environmentally acceptable diets?

With regard to biological monitoring, he said that there were a number of techniques which were being developed which could be used *in situ* in the presence of specific toxins, and the WRc was currently developing a technique on the scope for growth of the freshwater shrimp. That might well be applicable to the presence of chemicals used on fish farms.

Mr S. Ross (Clyde River Purification Board), with regard to consenting fish farm discharges and chemicals, said that one proposal of his Board in the monitoring of caged fish farms, which had previously been used for the monitoring of industrial discharges, was to moor cages of mussels close by and compare with a control site. The use of the 'scope for growth' technique, which had already been mentioned, was being considered.

With regard to biological indicators, he said that his Board had had a 'blanket' condition on all consents for many years, which was that the discharge should not be harmful to flora and fauna downstream. This had never been challenged in court. Another condition which had been put on land-based fish farms was that the discharge should not cause a visible growth of sewage fungus immediately downstream of the discharge. He felt that people should not be afraid to use these sorts of consent conditions.

Mr A. Reid (Forth River Purification Board) said that with reference to earlier comments concerning the setting up of a number of fish farms on a stretch of river, he wished to give assurances that when consent standards were set, account was taken of the legitimate use of downstream users. By carrying out a mass balance calculation it was usually fairly easy to assess the required consent conditions in order to protect the downstream water quality.

He asked Mr Saunders-Davies if he had any experience or knowledge of the use of copper sulphate in the treatment of eye fluke by killing snails?

Mr R. Allcock (Tay River Purification Board) said that one aspect that puzzled him was why the water authorities had not taken a tougher line on fish farming because one control that could be used with a consent was a volume limitation. This made it possible to limit volume even in the absence of abstraction licences, which are not issued in Scotland.

Mr R. Thomas (Welsh Water Authority), in a written contribution, commented that Mr Saunders-Davies had said that he did not consider that fish farm's discharges were a significant disease risk to receiving waters. Within West Wales there had recently been an objection to the establishment of a new fish farming venture 1.5 km above an existing business. The existing business had objected on the grounds of a significant risk of disease arising. He understood that a similar problem had arisen in the Wessex WA area and that the Wessex WA had concluded that fish disease organisms were embraced within the controls provided by COPA. He asked Mr Crewe if he could confirm that this was the situation, and if so, could he describe the treatment process that was envisaged to disinfect the fish farm discharge and detail the monitoring techniques that would be employed? He also asked Mr Saunders-Davies to comment on the Wessex WA approach to this problem.

PAPER 11

Mr R. G. Toms (UK Centre for Economics and Environmental Development), opening the discussion, said the paper referred to nitrates in drinking water, nutrients in rivers and lakes, and to protection zones. He said that controls over all these had a place, but he felt that each was separate and could not be interrelated. He did not believe that protection zones were going to be of much use in controlling nitrates in drinking water; one would presumably have to limit the amount of nitrate that the farmer used, and he questioned how this would be monitored. He said that if he was a farmer in the situation where he had been told to limit the use of nitrate fertilizer, he would press for the development of wheat or other crops genetically engineered so that they could use the nitrogen from the atmosphere and the yield could be maintained. Such a situation would lead to the same problem because much of the nitrate in water originated from the decomposition of plant residues left in the soil at the end of the growing season, and the quantity of nitrate would reflect increased crop yields, whether from artificial fertilizer or the improved use of naturally occurring material.

Use of such zones had a long time span, and he did not believe that the EC was willing to wait that long for the reduction of nitrate levels. It was also inefficient because one was reducing nitrates on vast quantities of water, most of which was not going to be used. He felt that nitrate removal from water supplies was most efficiently achieved artificially after abstraction. He said that this was already undertaken in other countries, although he agreed that more work was necessary before full-scale introduction in the UK.

With regard to nutrient control, he said that he could understand there being eutrophication problems with some of the long European rivers, but most of the UK rivers were relatively short and nutrients did not cause problems except in limited areas such as the Norfolk Broads or the Lake District. He asked if the UK should agree to an EC Directive to reduce eutrophication that was not needed in the UK?

He felt that the most important use for water protection zones was in areas where industry or farmers were storing dangerous chemicals upstream of water intake points. Improved control was needed as accidents would take place, and any spillage could pollute rivers and water supplies before one knew about it. He said that it was to this end that the use of protection zones would be of help by requiring greater control on the storage and use of these materials. However, he did not understand how they were going to help meet the nitrate conditions of the EC Directive within the foreseeable future.

In conclusion he said that the author had referred to the 'polluter pays' principle, but Mr Toms said that at the end of the day it was the consumer who paid, because the 'polluter' passed on the cost in his prices to the customer.

Mr L. B. Keith (Department of Agriculture and Fisheries for Scotland) said that the objective of the proposed legislation was to ensure that public water supplies complied with the EC Directive 80/778 limit for nitrate concentration of 50 mg/l. That would be in the interests of public health and achieved through the introduction of a scheme of measures aimed at controlling the entry to water supplies of nitrate from agricultural sources.

The author had commented that the Government would prefer agreements on a voluntary basis in the first instance, but would reserve the right to introduce compulsory powers. Since safeguarding public health was the ultimate aim of such powers, should the controls not be targeted at sensitive 'hotspots', with designation affording protection to the water supply through compulsory rather than voluntary agreements?

Mr H. Black (Farmer) said that there were some hundreds of thousands of houses about the UK which relied upon individual private water supplies, and he wondered how these were ever going to comply with the nitrate directive? Some authorities thought that they would have to condemn houses because of it, and others thought they would have to condemn the water sources and distribution systems. Others thought that they would have to designate their own water protection zones.

He said that there was supposed to have been a consultation paper on private drinking water produced by the DoE in association with the progress of the Water Bill through its committee stage. He said that the paper had not been issued. He hoped that this aspect would receive due consideration.

Mr K. Guiver (Southern Water) asked Dr Harryman to confirm that the vulnerable zones with regard to discharges from sewage-treatment works would apply only where there was a danger from the eutrophication limit being exceeded, as opposed to the 50 mg/l nitrate on drinking water.

He said that the proposal referred to 10 mg/l of nitrogen, and as far as sewage-treatment works were concerned he believed that was total nitrogen, which was what the EC may be interested in in terms of eutrophication. With regard to sewage effluents, that would be not only nitric nitrogen but also ammoniacal nitrogen—to get down to 10 mg/l as total N would be very difficult.

Finally he asked Dr Harryman what chance he thought there was during the consultation period of not making the water protection zones compulsory in the EC legislation.

Mr L. Woodward (Elm Farm Research Centre) said that it was known that modest restrictions on agricultural practice in areas of fertilizer application would not actually achieve the levels in reduction of nitrate leaching that were needed to be achieved. He said that protection zones would really be wholesale changes of land use which would have significant effects not only on farm incomes but also on communities in a water protection zone. That would be the situation unless a middle way could be found, and that was where he believed that organic farming should be considered as a preferred management option within environmentally sensitive areas. Basically, organic farming entailed restrictions on practice, and also positive benefits in terms of food production and environmental protection.

Mr H. Headworth (Southern Water) said that he disagreed with Mr Toms, and thought that there were merits in the creation of aquifer protection zones. He said that Dr Harryman envisaged that initially it was likely that the NRA would be submitting its proposals to public inquiries where people could legitimately make representation. He believed that these would be tortuous and difficult to set up, and he wondered if there was merit in the DoE co-ordinating test cases for selected areas to expose and address the issues surrounding each particular area, in the hope that subsequent sizeable inquiries might be avoided. This would enable the NRA to get on with its job properly without always fearing protracted public inquiries.

He said that the EC confused vulnerability with protection zones. He said that vulnerability was the assessment of risk, whereas a protection zone was the specification of controls. Dr Harryman's paper suggested that the whole of a catchment could be subject to designation, and he asked how the DoE envisaged this overall wide-scale designation of vulnerability zones tied in with the more localized specification of protection zones, each of which might require an inquiry? He could not see the two aspects matching.

Mr H. Gammp (Sandoz Ltd) said that he had spoken to people in Germany, and there seemed to be a belief that the creation of protection areas was also a means to reduce over-production within the EC countries. He said they thought that at the moment one could not directly reduce over-production for political reasons. However, based on the environmental issue it was acceptable. He asked Dr Harryman to comment on this aspect.

Mr C. H. Mathias (Laurence Gould Consultants Ltd) asked if restrictions were likely to be introduced on a catchment-wide basis, or would they be more localized? The reason for this question was that the two alternatives were likely to result in considerable differences in the degree of land use change and/or intensity of farming practices required, as well as having social impacts.

Dr H. A. C. Montgomery (Consultant), in a written contribution, noted that there was a proposed EC Directive on protection against pollution by nitrate from diffuse sources. He hoped that there would be a searching enquiry before any waters were designated as vulnerable to eutrophication by nitrate. He had taken part in a study of Langstone Harbour where it had been thought that it would be necessary to denitrify the effluent from Budds Farm sewage works at Havant, to prevent the Harbour from suffering catastrophic growths of benthic algae. However, it was shown, after a long and detailed investigation, that the habitat would in any case be unlikely to accumulate benthic algae in quantities much greater than those already occurring in the 1970s, which were acceptable, and that drastic and expensive measures at Budds Farm were therefore not required (Soulsby, P. G., Lowthion, D., Houston, M. and Montgomery, H. A. C. (1985) The role of sewage effluent in the accumulation of macroalgal mats on intertidal mudflats in two basins in Southern England. *Netherlands J. Sea Res., 19*, 257–263).

In many fresh waters, phosphate was likely to be more important than nitrate in promoting algal growth. Dr Montgomery hoped that administrators and politicians would not jump to conclusions as to which waters were vulnerable to unacceptable eutrophication by nitrate.

Authors' replies to discussion on Papers 8, 9, 10, and 11

PAPER 8

In response to Mr Binnie, **Dr Bailey** said that he was familiar with the Silsoe College computer program and had helped in its development. He said that it was very similar to the one which he had used, but it went one step further in that it translated the irrigation requirement into a yield response which was extremely useful for assessing benefits in an economic analysis. He said that it was not widely available, however, being still in the development stage and at present could be used only for the sugar beet crop.

He said that there were a number of computer programs from various organizations that were used in day to day scheduling, for example ADAS.

With respect to the normal year irrigation usage, he said that this fitted in well with his calculations for requirement. He quoted figures for the years 1982 and 1984 for which he said there were good census data. He regarded 1984 as having been a dry summer and the usage had been 98 million m^3—far below the licensed amount. In 1982 the amount used was only 55 million m^3. This demonstrated that the licences were well in excess of the volume actually being used. Even in 1976, because of the practical restraints of irrigating, (for example setting up and moving equipment), the amounts used in many cases were much less than the licensed quota.

Answering Mr Oxley, he said that the irrigation amounts that had been used were the amounts required to prevent each crop coming under any stress, which would affect either yield or quality. This would determine the date within the program on which water was applied. The amounts of water to be applied were carefully chosen so that they did not in any case return the soil to field capacity, which would be subject to leaching risk if there was subsequent rainfall. He said that there was no major concern in the irrigated areas of the UK with regard to build-up of salts in the upper soil layers because winter rainfall was usually adequate to leach them down. In fact, salt was used as a fertilizer with sugar beet.

He accepted that the use of the fifth driest year wasn't always the best approach, and in fact if he was carrying out an economic investment appraisal on a particular farm, he would not necessarily use the fifth driest year but would consider each one of the twenty years. It was, however, the conventional method of assessment of irrigation requirement, and for this reason had been adopted in this analysis.

In reply to Mr Thompson, he said that in carrying out the analysis for any particular farm, site surveys would be carried out to assess aspects such as levels of rooting. It would have some effect on the result, but this was unlikely to be substantial except in the wetter years. For such a broad national assessment as had been made for this paper, it was necessary to assume a rooting depth for each crop, and previously published figures had been chosen. Although there could be areas where the rooting depths would be less, there would also be areas where they were greater—the assumption had been made that these would cancel each other out.

Responding to Mr Franklin, the author said that he had divided the country into seven areas for this purpose. He said that the Meteorological Office and MAFF published a book called the *Agroclimate of England and Wales*, which divided the country into 52 areas, each of which was described in detail for many weather parameters that were regularly measured. One of the key measurements in soil irrigation was soil moisture deficit. Of the 52 areas in the country, consideration had been given to the maximum soil moisture deficit attained during the summer months.

By using that particular figure it had been possible to divide the country into seven clear groups, which he considered was adequate for the sort of analysis that had been carried out. The figures should not be taken out of context, and, for example, applied to an individual farm, but they were acceptable for a regional view.

Further to Dr Cook's query, Dr Bailey said that interception losses from the crop canopy were small and had been ignored. In a typical irrigation with 25 mm water, the canopy would not intercept more than 1–1.5 mm.

In answer to Mr Adlam he said that he had no information for overseas, but in the UK it depended much on location and soil type. For example, on Nottinghamshire sands there was no need for artificial drainage systems—the natural drainage in the sandy soil was enough. On the other hand, on the silts in Lincolnshire drainage was very important.

Replying to Mr Svoboda, Dr Bailey said that there were various impurities that could make water unsuitable for irrigation purposes. He said that there were published limits which should not be exceeded, but these had been derived from work that had been carried out in Canada, where the transpiration demand in the summer was much higher and more irrigation was applied. The use of these limits was likely to be rather cautious. A good example was shown by chloride. According to published standards, when irrigating potatoes one should not allow the chloride level to exceed 600 mg/l. Experiments in this country have now shown, however, that while significant concentrations of chloride will depress yields, use of irrigation water containing 2000 mg/l had produced higher yields than leaving the crop unirrigated!

In answer to Mr Morris he said that inclusion of economic aspects would reduce the amount that was applied to some crops such as cereals. He said that it would have the effect of reducing the irrigation amounts estimated and would strengthen his case even further.

Replying to Mr Burroughes, Dr Bailey said that the National Rivers Authority did have legal power to revoke or vary a licence if it considered it appropriate to do so. Such power had always been available to the water authorities, but had not been used significantly. This might be partly due to the fact that the authority could be liable to pay compensation if the farmer could show loss as a direct result of that action. Thus, to protect themselves from legal action, the authority would need to be absolutely certain of an individual farmer's requirements before varying his licence.

A further point was also relevant here. The reduction of a licensed quantity in one location did not necessarily imply that an equal quantity could then become available at or near that location or elsewhere, as the constraints on licensing varied from place to place. Every new licence proposal had to be considered on its merits and was subject to the primary legal constraints that existing abstraction rights must be protected. In a situation where an unused licence was the only constraint on the grant of a new licence to another party, the authority would have to give careful consideration to whether action to revoke or vary the constraining licence would be justified.

PAPERS 9 AND 10

Mr Saunders-Davies, in response to Mr Smith, said that he was aware of the discussions that had been taking place in Scotland. He said that perhaps what he had intended to say was that a practical means of applying consent conditions to non-flowing discharges was difficult to conceive. How did one actually measure a discharge where there was no flow, and did one enforce the standard? He asked Mr Smith how he would set about defining parameters that were measurable and hence enforceable for a non-flowing water fish farm discharge?

In response to Mr Toms, **Mr Crewe** said that he agreed with much of what he had said, and the use of biology in rivers had been found to be very useful. It indicated what was going on all the time rather than at the instantaneous point in time that a sample was taken for chemical analysis. In consideration of biological indicators for the protection of supply intakes, Mr Crewe said that this was a question which had received much thought. The question was how did one do it? The use of fish had been considered by measuring their response to water pumped through a tank containing them. The problem was that the fish tended to respond to other things as well and was, therefore, not as reliable as one would wish. He felt that it would be very useful if a biological system could be developed for assessing water quality.

Mr Saunders-Davies said that he felt that the use of population equivalents could be extremely misleading, particularly if they were based, as he understood, on 1982 figures. There had been a vast improvement since then—from 1300 kg/tonne of fish to about 500 kg/tonne of fish. A question which he felt had not been brought up was the absorptive capacity of the receiving river. He did not suggest that fish farming had a right to use all the regenerative capacity of the system, but he felt that this was an issue which needed to be addressed within the whole concept of the use of water for fish farming and discharge of other biological wastes into a water system.

He agreed that in principle the use of biological indices, rather than chemical, was attractive. However, he suspected that from the point of view of enforcement they simply could not be used. One needed something that could be measured with some degree of certainty in order to take some steps to respond to it, although biological monitoring of waste water systems must be worthwhile.

He drew attention to a Freshwater Biological Association report on the Hampshire Avon, which he said, by and large, did not find fish farming as responsible as some people had expected.

In response to Mr Walls, Mr Saunders-Davies said that he would be prepared to provide information on chemicals. He stressed that the fish farming industry did not use hormones in the production of fish for consumption. He said that extremely small quantities of hormones had been used in hatchery situations. He said that perhaps it was a lack of knowledge that led to the slightly emotive concern about chemicals. Most of the chemicals were of a simple nature and were not pesticides. Antibiotics were used under veterinary prescription, but these were in the food, not in the water. Calculations had been carried out a few years ago, he thought based on the Hampshire Avon, on the theoretical premise that every fish in the system was treated with a prescribed dose of antibiotics, and that every antibiotic totally missed the fish and was absorbed by the water system, resulting in virtually undetectable quantities in the water. He said that it should be noted that if there was not a deleterious effect on the fish in a farm from the use of chemicals, then there should not be a problem in the water system, particularly bearing in mind the further dilution and absorption downstream.

Replying to Mr Wade, Mr Crewe thanked him for his comments and said that he would like more information on such systems.

In answer to Mr Seager, Mr Crewe said that with any type of discharge the water quality team in the Wessex WA were constantly in touch with the discharger and advising of the latest developments, especially where that could improve the quality of discharge. He said that he did not know specifically whether there was encouragement to use better diet foods on fish farms.

Mr Saunders-Davies said that fish farming, like any other agricultural activity, had an obvious interest in improving efficiency, and he felt that such encouragement would be self-generating. The improved diets which had higher digestibilities led to more economic production, and adoption was already widespread. Perhaps one of the problems with the former diets had been the amount of organic filler needed to achieve good pelleting quality, etc, to make the diets work mechanically. This had resulted in indigestible and unusable products as far as the fish were concerned being incorporated into the diets for manufacturing reasons. The tendency with expanded rather than pelleted diets, and the better understanding of nutrition and digestion, had resulted in a great reduction in this. He said that the only negative aspect was that fish farming worldwide had now become such a successful and large industry that it was a very significant user of the available high quality fish-sourced proteins, and a consequence of this might be that in the future the industry would have to look at how to use other proteins which, in their present form, were less usable and less digestible by the fish. Research would have to be directed toward ensuring that some species could perhaps use vegetable derived proteins.

Mr Saunders-Davies said that the argument about consents being a basis for 'licensing' volumes was a view which he took. He believed that the water industry had to address the problem and come to a conclusion.

In response to Mr Reid, he said that he had no experience in the use of copper sulphate for eye fluke, but he said that years ago copper sulphate was a readily accepted treatment for gill disease and fin rot. It was effective and fairly safe in high hardness waters, but he believed it was a hazardous experience in the soft waters of Scotland.

PAPER 11

Replying to the discussion, **Dr Harryman** said that with regard to nitrate in drinking water, a protection zone would not help in the short term. However, in the longer term, protection of sources would lead to improvements in water quality. He agreed that the immediate need was for treatment or blending after abstraction; this requirement would continue for a number of years. Protection was a long-term measure: for example some aquifers needed 40 years to respond to measures taken on the surface before there was improvement in water quality.

Dr Harryman said that he saw protection zones as helping to control the amount of nitrate going into the water, albeit in some cases in the very long term. He said that the type of measures that would be introduced within the zones would vary from one to another, but would, he believed, include agricultural restrictions in many cases. The approach to policing of such restrictions would obviously depend on the type of restrictions. If the restriction was that the farm was being taken out of production, then compliance would be relatively easy to check. If there were to be limitations on the amount of nitrate added to land, then obviously that was more difficult to monitor.

As a member of the European Community, the UK would be a party to negotiation on the Nitrate Directive. In the negotiations the UK would be seeking changes to the proposals to reflect the situation in this country. As far as the Commission was concerned, the Nitrate Directive and the Waste Water Directive would be two parts of a whole which would be used to control nutrient input across Europe.

Chapter 4

The final chapter of the Symposium discusses environmental protection. The contributions span the impact of land drainage from an agricultural benefit and environmental impact standpoints and outline the ways in which drainage activities are being modified to meet environmental criteria. Land drainage schemes invariably end up in improvements to rivers and the corridors of land through which they flow. These areas are one of Britain's major wildlife resources. Changing patterns of river low can result in changes to the nature and extent of river habitats. In extreme cases, the linking of these river corridors have isolated some habitats by intensification of land use.

Intensification of land use has led to an inevitable over-production where the market forces of supply and demand are no longer in harmony. Some land must be transferred into other uses over the next decade to redress this imbalance. European legislation provides the basis for opportunities for the farmer to explore set-aside policies such as extensification in cereals and beef, farm diversification, support for afforestation and woodlands. The potentials of these options are discussed, together with the effect that a structured change in land use will have on water quality and the improvements to the environment and conservation.

12

Land drainage: agricultural benefits and environmental impacts

J. Morris [†]

ABSTRACT

The paper reviews the context and contribution of agricultural land drainage in Britain and discusses the design and performance of land drainage improvement and maintenance schemes. The environmental impacts of land drainage are considered, and how drainage activities can and are being modified to meet environmental criteria.

KEY WORDS   Agricultural drainage   Economics   Cost benefit   Environmental impact

INTRODUCTION AND SCOPE

Government incentives for land drainage improvements such as river flood alleviation, arterial drainage schemes, and field drainage installations, have been an important component of agricultural support in Britain. More recently, over supply of agricultural commodities, pressure on public funds, concern with the environmental impacts of intensive agriculture, and competition from non-agricultural land uses have questioned the desirability or indeed feasibility of agricultural systems in Britain which do not protect environmental quality. The environmental impact of land drainage for agriculture has been central to these issues. This paper reviews the context and contribution of agricultural land drainage in Britain and discusses the design and performance of land drainage improvement and maintenance schemes. The environmental impacts of land drainage are considered, and how drainage activities can and are being modified to meet environmental criteria.

AGRICULTURAL DRAINAGE OBJECTIVES

Water is essential for agriculture, but too much of it can limit crop growth and cropping activities. The basic aim of agricultural drainage is to alleviate or prevent the problems caused by excess water in order to facilitate more productive agriculture.

[†]   Senior Lecturer in Agricultural Economics and Management, Department of Rural Land Use, Silsoe College, Silsoe, Beds.

The agricultural drainage problem may be due to coastal or fluvial flooding, and/or the presence of seasonally or permanently high water tables. Poor drainage due to flood inundation often calls for river re-channelling and embanking works. Where poor drainage is due to waterlogging, the preoccupation is improving arterial watercourses, installing piped under-drainage, and other measures such as subsoiling which improve the movement of water in the soil.

Figure 12.1 summarizes the principle benefit of improved drainage for grass and arable enterprises. They can be generalized in terms of improvements in the soil/plant/ water growing environment and in the 'trafficability' of soils which allow more timely access for field activities, together with a reduction in crop damage caused by inundation.

A review of relevant agricultural drainage literature is contained in Morris *et al.* [1]. Much research into crop responses to drainage has measured soil wetness in terms of depth to water table and has related yields to this parameter [2, 3, 4]. Experimental results show a wide difference in the tolerance of crops to the timing and duration of waterlogging or inundation [5, 6]. Recent UK research has examined the yield effects of waterlogging at different stages of crop development, concluding that water table control to 0.5 m is adequate provided that incursions above this level are brief [7]. With respect to field trafficability and work day probability, soil strength and load bearing capacity vary inversely with water content and water table height [8, 9, 10]. Better drainage reduces the risk of untimeliness yield penalties for critical operations such as seed bed preparation or fertilizing [11, 12, 13]. In many respects requirements for trafficability and workability may impose stricter conditions than the crop itself. In grassland systems, improved drainage can reduce the risk of surface damage or 'poaching', allow earlier access for grazing stock and the timely application of nitrogen fertilizer, and facilitate a move to more intensive and predictable silage making systems [14].

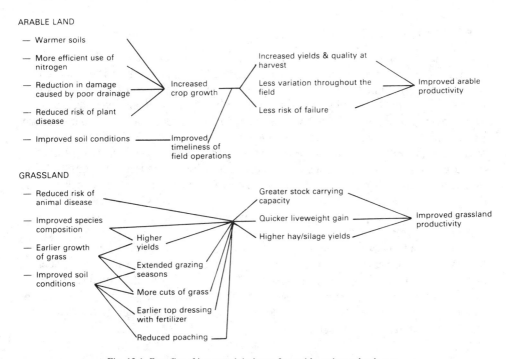

Fig. 12.1. Benefits of improved drainage for arable and grassland uses.

DRAINAGE METHODS

The purpose of drainage design is to control the level of the water table and the risk of flooding according to limits required for agricultural production. River, arterial, and field drainage in the agricultural catchment must be considered as a whole.

Field drainage, which remains at the discretion of the private farmer, is mainly a traditional art based on experience of local soils and conditions. Open ditches may adequately serve soils which are free draining, otherwise tube drains or inclined mole drains may be required, with or without permeable stone fill. The scientific approach to drainage design attempts to control the water table by finding suitable drain spacing and depth for the expected rainfall and hydraulic characteristics of the soil, using mathematical equations and modelling. Agronomic trials provide guidelines for preferred water table heights, which for most crops average 50 cm below the surface. Tube drains are often laid at 0.8 to 1.5 m, depending on soil type and field conditions with about 0.15 m freeboard at the outfall. Mole drains are placed at about 0.5 to 0.6 m depth [15]. With respect to flood risk, practice has provided criteria for arterial designs to give minimum acceptable 'whole year' flood frequencies. This is 1 in 2 to 1 in 5 for most field crops, and 1 in 10 years for high value horticultural crops. Desired standards for flood risk during the April–October period are more stringent [16].

In principle, the criteria and methods for drainage can of course be used to create and retain wet conditions for natural habitats if so required.

THE BENEFITS OF FARM DRAINAGE

For the farm business, the benefits of land drainage may occur in a number of ways [17, 18, 19, 20, 21]. Drainage may allow the *reclamation* of land previously unsuited to commercial agriculture. Improved drainage could allow the *intensification* of existing enterprises, particularly arising from the use of yield increasing inputs and practices, such as improved forage conservation. Alternatively, better drainage could facilitate *land use change* and a switch to new, more profitable enterprises, such as a move from permanent pasture to an arable/ley rotation. Furthermore, *reduced production costs* could result owing to improved working conditions or savings in forage conservation costs due to extended grazing seasons. Where drainage incorporates a significant part of the farm, there may be *benefits for activities on other parts of the farm* such as where increased stocking on lowland pastures may release upland areas for arable production whilst maintaining herd size.

Figure 12.2 provides a basic framework for agricultural drainage benefit assessment. Improved land drainage offers a degree of flood alleviation and/or an improvement in outfall levels in the arterial drainage system. The extra value of intensifying existing, or moving to new farming systems, can be expressed in terms of changes in enterprise gross margin and farm-level fixed costs. Methods for appraising land drainage improvements are described elsewhere [22, 23].

THE VALUE AND PERFORMANCE OF AGRICULTURAL LAND DRAINAGE

During the 1970s there were increased demands for rigour in the design and appraisal of agricultural land drainage schemes. Schemes were criticised for their somewhat arbitrary and optimistic estimates of benefits and inadequately detailed design and cost estimates. As Britain moved towards self-sufficiency and surplus in indigenous foods, appraisals were criticized for not adjusting benefit estimates to reflect either direct or hidden government subsidies to farmers. Furthermore, there was growing concern that whilst agricultural benefits might be overvalued, environmental impacts received insufficient attention either at the design or analysis stage.

In this context, and with a view to improving the assessment of agricultural drainage benefits, Silsoe College estimated the value to farmers of changes in drainage regimes on 22 Water Authority and Internal Drainage Board schemes in England, covering over 300 farmers and 10 000 ha [24]. At farm level, the greatest benefits (Table 12.1) were associated with land use change to higher value cropping, and the installation of field drains. In the case of grassland, benefits were particularly associated with increase in nitrogen use, and changes in grass conservation for winter feed. Key factors influencing farmer uptake of benefits were the extent to which poor drainage restricted desired land use, security of tenure, farm type and size, and recent changes in farm management.

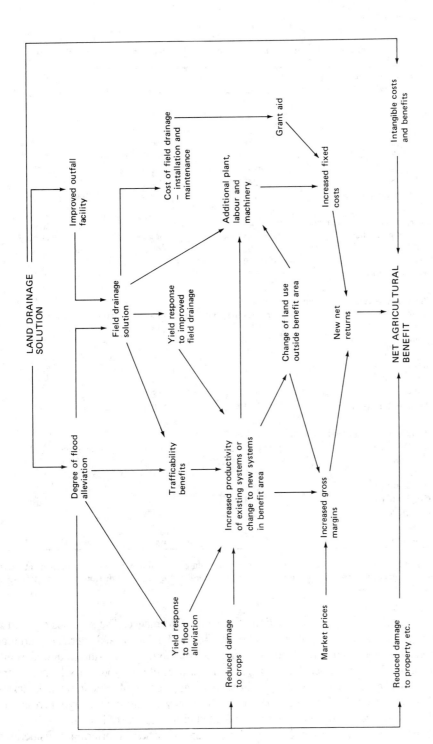

Fig. 12.2. Determination of drainage benefits at farm level.

Table 12.1. Agricultural benefits of land use change, field drainage and grassland management

Benefits of land use change

Shifts in land use[a]	Extra net revenue (£/ha)	% of area	Extra net revenue (%)
−1	73	2	3
0	92	59	37
1	195	13	17
2	188	6	7
3	285	18	35
4	195	1	1
5	243	1	2
Average or total	150	100	100

Benefits of field drainage installation

Drained post scheme	Extra net revenue (£/ha)	% of area	Extra net revenue (%)
No	101	60	33
Yes	199	40	67
Average or total	150	100	100

Benefits of increased use of nitrogen on grass

Additional N (kg N/ha)	Extra net revenue (£/ha)	% of area	Extra net revenue (%)
None	25	38	18
1–49	21	13	5
50–99	47	25	22
100–149	91	13	23
150–199	164	3	10
200 +	138	8	22
Average or total	52	100	100

Benefits of change in grass use or conservation

Change in grass use	Extra net revenue (£/ha)	% of area	Extra net revenue (%)
No change	40	71	56
Increased cuts	51	1	1
Hay to silage	67	15	19
Grazing to hay	53	1	1
Grazing to silage	59	3	4
Hay to grazing	104	9	19
Average or total	52	100	100

[a] Shift indicates numerical change in land use where: 0 = unused, 1 = rough grazing, 2 = permanent grass. 3 = temporary grass, 4 = crop/ley rotation, 5 = crops, 6 = horticulture. E.g., a shift from permanent grass to crops = 5 − 2, i.e. 3.

The study also allowed an evaluation of scheme performance. The schemes covered a broad range of sizes and situations, including river embankment and regrading, arterial improvements, and pumping. Benefits over a 30 year project life (this required some extrapolation of observed benefits over remaining project life for younger schemes) were set against scheme costs. Fourteen of the 22 schemes appeared profitable in financial terms at the Treasury's 5% discount rate. The poor performers were those with large increases in capital costs beyond those predicted, or very limited farmer uptake of benefits.

The best performing schemes were large pumped schemes offering considerable economies in capital and running costs, where flooding problems had been virtually eliminated and improved arterial networks had removed the drainage constraint. Here previously wet areas had been assimilated into the rest of the farms' mainly arable cropping system. Without exception, the agriculturally beneficial and worthwhile schemes contained informal or formal farmer drainage organizations which were instrumental in pressuring for drainage improvement, and pushing for follow-up work.

The evaluation showed that, during what was a period of national expansion in agricultural output, land drainage investments were financially attractive where poor drainage was a real constraint on commercial farming. The 'reality' of this constraint was usually voiced by farmers themselves. In other cases, particularly where farms were small and fragmented, and followed traditional methods, benefit uptake and financial performance of schemes were relatively low.

RIVER MAINTENANCE EVALUATION

In present circumstances, there is little need or justification for new agricultural drainage schemes. Attention has now turned to the scrutiny of land drainage maintenance programmes, both in terms of improved programming of physical works [25] and the application of management criteria to justify levels of service and prioritize maintenance work [26]. With respect to agricultural criteria, Silsoe College have [27], in collaboration with Severn–Trent Water, developed a simplified model (Fig. 12.3) for evaluating river maintenance expenditure which combines basic data on agro-climate, soil types, topographical and catchment characteristics, field drainage, flood risk, and land use, to identify arterial drainage design parameters. On the basis of research evidence and field observations, drainage productivity classes have been defined in terms of water table depths during the critical spring period. Water tables of less than 30 cm depth indicate a 'breakdown' situation, between 30 cm and 50 cm a 'low' productivity class, and greater than 50 cm a 'normal' drainage situation. The method estimates field water table depth on the basis of freeboard in the adjacent watercourse for specified soil hydraulic conductivity and flood plain topography.

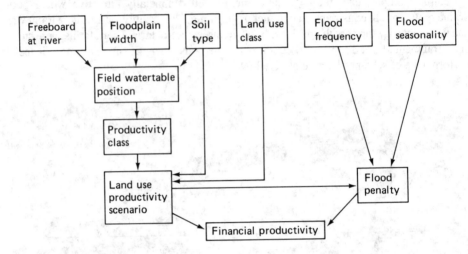

Fig. 12.3. An approach to river maintenance for agriculture.

Where the water table is controlled by underdrainage the desired freeboard is determined by drain depth in the field as influenced by soil type, drainage problem, and pipe run and gradient. Productivity class is normal provided that freeboard is sufficient to avoid submergence of outfalls. Otherwise productivity classes could be low or break down.

Where sites drain naturally to the river, drainage equations can relate water table height to freeboard (hydraulic head), seepage rate (hydraulic conductivity), land slope, and rainfall. For specified local conditions, freeboard is used to estimate water table height and thereby a drainage productivity class.

Surveys in the catchment area can be used to compile scenarios for each land use class/productivity level combination. Each is described in terms of financial performance reflecting the impact of waterlogging and flood risk. These scenarios are then used to show the benefit associated with maintaining a given level of service, the likely 'disbenefit' associated with lower standards, and the justification for maintenance expenditure. The method is being refined to allow for variable rainfall conditions, floodplain ditch layouts, and type and seasonality of maintenance operations, and extended to incorporate the impact of maintenance on environmental quality.

THE ENVIRONMENTAL IMPACTS OF LAND DRAINAGE

Land drainage may have a number of environmental consequences for natural vegetation, wildlife, groundwater levels and quality, and catchment hydrology. Concern for the impacts of land drainage on the environment have focused on the permanent loss of wetland sites and the effects of changes in agricultural practices after drainage.

Natural vegetation can be affected in two main ways [28], where the expansion of agricultural production displaces natural plant communities, and where lower tables result in a change in species composition in previously wetland sites. Wildlife resident in or indirectly dependent on wetlands, can be depleted by the drainage, change of use, and consequent loss of habitat [28, 30, 31]. Depending on local conditions, land drainage may, by diverting water into rivers, reduce the recharge of the aquifer and cause a lowering of groundwater levels. This can result in the drying up of ponds, reduced summer stream flows, and peat shrinkage. In addition, changes in land use associated with field and arterial drainage may lead to modifications in the hydrological cycle. Whereas field drainage, by quickly drying out soils, tends to increase the storage capacity and reduce run-off and flood risk during heavy storms, improved arterial drains tend to create earlier peak flows in the main channels and thereby increase flood risk [32, 33, 34]. River works, including regular maintenance, can disrupt the hydrological and biological balance of the river, with consequences for fish and other aquatic creatures [35, 36].

The alteration of the landscape and ecology of a site can have a major impact on its use for recreation and amenity. Anglers, bird watchers, outdoor sports persons, and those in pursuit of rural tranquillity may be affected because of land drainage.

INCORPORATING ENVIRONMENTAL CRITERIA INTO DRAINAGE MANAGEMENT

In response to environmental concern, a number of guidelines were drawn up to reconcile drainage and environmental needs [37, 38, 39]. Responsibilities for conserving or enhancing environmental features, and for prior consultation with the Nature Conservancy Council, were embodied in the 1981 Wildlife and Countryside Act. More recently operational guidelines have been compiled for environmentally sensitive rivers and land drainage management [40, 41, 42], and for field drainage [43]. In addition, methodologies for assessing likely environmental impact are being developed and applied [44, 45]. Generalized environmental impact statements were required by MAFF in 1985 for grant aided land drainage schemes [46], and environmental assessments are now required for significant agricultural water management projects under the European Community Directive 85/337. More rigorous assessment of agricultural benefits is also likely to reduce the optimism of many drainage proposals [22, 23].

A most significant agriculture/environment initiative has been the designation of 19 Environmentally Sensitive Areas within which farmers are given financial incentives to farm in accordance with environmental priorities. A number of locations, namely the Somerset Levels, the Suffolk Rivers, the Broadlands, are sites where some form of managed drainage sustains a valued complex of traditional farming, ecology and landscape. Here there are opportunities for continuing or developing drainage management strategies which reconcile agricultural and environmental interests. In many respects the data, techniques, and expertise developed for designing agricultural drainage are applicable in principle to designing measures to protect or create wetland habitats. For instance, much research effort has examined soil/plant/water relationships which have been used in the design of agricultural drainage schemes. The same approach could be used to describe water regimes required for natural wetland species. The sensitivity of natural species to changes in soil water regimes can be assessed, particularly with respect to conditions preferred for agricultural species. In the same way that engineers have created an environment for commercial agriculture, so can they be encouraged to develop engineering and management methods to create or protect the desired water regimes for 'natural' habitats. By way of example, Silsoe College examined the feasibility of wetland creation/retention in general drained areas for CIRIA [47]. More recently the College is examining the influence of river and arterial drain levels on ditch-side and field soil water conditions as they influence habitats for natural species [48]. The results have some interesting implications for site management, and point to considerable scope for reconciling farming and wildlife interests within a traditionally managed agricultural system. In clayey valley bottoms in North Wales, river maintenance, in the absence of field drainage, often has limited influence on surrounding water tables. This observation, together with a closer examination of agricultural drainage needs, has recommended less frequent maintenance, concomitant ecological benefits [49]. These studies suggest that the hydrological, engineering, and agronomic principles and methods used to design drainage for agriculture are relevant for the design and management of water regimes in areas where priority is given to conservation, recreation, or amenity. Furthermore, selected aspects of cost benefit methods can be used to assess the value of wetland conservation, not only in terms of agricultural output forgone, but also the value of (non-priced) services flowing from these sites [50].

CONCLUSIONS

Agricultural drainage has made, and will continue to make, a major contribution to Britain's farming. The efficient operation and maintenance of land drainage systems will, if anything, increase in importance for many commercial farmers facing diminishing profit margins. Given changing agricultural and environmental priorities, environmental criteria will play a major part in future land drainage improvement and maintenance activities. Initiatives by environmental groups, farmers, and Government are helping to reconcile different interests, especially in areas where many valued environmental qualities are sustained by a managed, albeit traditional farming system. The recent policy initiatives on farm diversification and extensification should provide further scope for matching economic and environmental criteria. In this context, a greater understanding of the interrelationship between water regime requirements for agricultural and environmental management can only be beneficial, and an opportunity for environmental and agricultural specialists to work together.

REFERENCES

[1] Morris, J., Hess, T. M., Ryan, A. M. & Leeds-Harrison, P. B. (1984) *Drainage Benefits and Farmer Uptake*. Report to Severn–Trent Water Authority, 4 volumes, Silsoe College, Silsoe, Bedford, MK45 4DT, UK.

[2] Visser, J. H. (1958) *De Landbouwwaterhuishouding in Netherland.* Rapp. Comm. Onderz. LandouwWaterhuish.ned.TND. No. 1.231.

[3] Sieben, W. H. (1964) *Het verband tussen onwatering en opbrengst bij de jonge Zavalgronden in de Noordoostpolder* (Van Zee tot Land, no. 40), Zwolle, Netherlands: Willink.

[4] Scraggs, R. W. (1975) Water management model for high watertable soils. *Am, Soc. Ag. Eng.* paper 75/2524, A.S.A.E., St. Joseph, Michigan.

[5] Trafford, B. D. (1972) *The evidence in literature for increased yield due to field drainage.* Technical Bulletin, Field Drainage Experiment Unit, ADAS, No. 72/5.

[6] Trafford, B. D. (1974) *Soil water regimes—what is known, the work which is in hand and suggestions for progress.* Technical Bulletin, Field Drainage Experiment Unit, ADAS, No. 74/13.

[7] Cannell, R. W. & Belford, R. K. (1982) Crop growth after transient water logging: advances in drainage. In: *Proc. of A.S.A.E National Drainage Symposium,* 4th Chicago, 163–170.

[8] Goodwin, R. J. & Spoor, G. (1977) Soil factors influencing work days. *Agric. Eng.,* **32**, 87–90.

[9] Spoor, G. & Goodwin, R. J. (1979) Soil deformation and sheer strength characteristics of sme clay soils at different moisture contents. *J. Soil. Sci.,* **30**, 483–498.

[10] Jarvis, R. H, (1977) The effects of timeliness of soil-engaging operations on crop yield. *Agric. Eng.,* **32**, 84–86.

[11] Smith, L. P. (1972) The effect of weather, drainage efficiency and duration of spring cultivations on barley yields in England. *Outlook Outl. Agric.,* **7**, 79–83.

[12] Armstrong, A. C. (1977) Field drainage and field work days: results from a national experiment. *Agric. Eng.,* **32**, 93–94.

[13] Wendte, L. W., Drablos, C. J. W. & Lembke, W. D. (1978) The timeliness benefit of subsurface drainage. *Trans. A.S.A.E.,* 21, St Joseph, Michigan, 484–488.

[14] Parker, A. (1983) *Benefits and economics of grassland drainage.* Ministry of Agriculture, Fisheries and Food, Land and Water Services Report RD/FE/14.

[15] Agricultural Development and Advisory Service (1982) *The design of field drainage pipe systems.* Ref. book 345, HMSO, London.

[16] Agricultural Development and Advisory Service (1979) *Getting down to drainage.* Arterial drainage and agriculture. Advisory Leaflet 736, Ministy of Agriculture, Fisheries and Food, London.

[17] Trafford, B. D. (1971) The Langabeare drainage experiment. *Agriculture,* **78** (7), July 1971, 305–311.

[18] Hunter, J. M. & Trafford, B. D. (1979) An economic appraisal of six years results from a drainage experiment on clay land. *Experimental Husbandry,* **35**.

[19] Bailey, A. D. (1979) Benefits from drainage of clay soils. *Paper No. 79–2549, Am. Soc. Ag. Eng.,* St Joseph, Michigan.

[20] Morris, J. & Calvert, J. (1976) Evaluating returns from drainage. *ADAS Quarterly Review.*

[21] Morris, J. & Hess, T. M. (1986) Farmer uptake of agricultural land drainage benefits. *Environment and Planning A,* **18**, 1649–1664.

[22] Hess, T. M. & Morris, J. (1985) *A computer model for agricultural land drainage scheme appraisal.* Ministry of Agriculture, Fisheries and Food, River Engineers Annual Conference, Cranfield, July 1985.

[23] Morris, J. & Hess, T. M. (1988) Agricultural flood alleviation benefit assessment: a case study. *Journal of Agricultural Economics*, (39), **3**, 402–412.

[24] Morris, J., Black, D. E. & Hess, T. M. (1987) *Drainage benefits and farmer uptake: a regional study.* Report to the Ministry of Agriculture, Fisheries and Food, Silsoe College, Bedford.

[25] Birks, C. J. (1985) *Planning river maintenance.* Ministry of Agriculture, Fisheries and Food, River Engineers Annual Conference, Cranfield, July 1985.

[26] Fitzsimons, J. & Chatterton, J. B. (1985) *The use of benefit: cost techniques in developing objective land drainage maintenance programmes.* Ministry of Agriculture, Fisheries and Food, River Engineers Annual Conference, Cranfield, July 1985.

[27] Hess, T. M., Leeds-Harrison, P. B. & Morris, J. (1986) *Evaluation of river maintenance for improving the agricultural productivity of riparian land.* Technical Paper, Silsoe College, Bedford.

[28] Hill, A. R. (1976) The environmental impact of land drainage. *J. of Environmental Management*, **4**, 251–274.

[29] Mountford, J. O. & Sheail, J. (1984) Effects of drainage on natural vegetation. In: Jenkins, D. (Ed.) *Agriculture and the Environment,* Proceedings of the ITE Symposium, Inst. of Terrestial Ecology, Cambridge, 98–101.

[30] Hellawell, J. M. (1988) River regulation and nature conservation. In: *Regulated rivers: research and management.* Wiley, Vol. 2, 425–443.

[31] Williams, G. & Bowers, J. K. (1987) Land drainage and birds in England and Wales. *Royal Scoiety for the Protection of Birds Conservation Review*, No. 1, Sandy, Beds., 25–30.

[32] Bailey, A. D. & Bree, T. (1981) Effect of improved land drainage on river flood flows. *Inst. of Civ. Eng. Flood Studies Report—Five years on.* Thomas Telford, pp. 131–142.

[33] Newson, M. D. & Robinson, M. (1983) Effects of agricultural drainage on streamflow case studies in mid-Wales. *Journal of Environmental Management*, **17**, 333–348.

[34] Robinson, M. (1987) Agricultural drainage can alter catchment flood frequency. *British Hydrological Symposium*, 6.1–6.10.

[35] Toms, R. (1975) The environment impact of land drainage work. In: *Conservation and Land Drainage*, Water Space Amenity Commission, pp. 6–15.

[36] Mann, R. H. K. (1988) Fish and fisheries of regulated rivers in the UK. In: *Regulated rivers: research and management.* Wiley, Vol. 2, 411–424.

[37] Miers, R. H. (1975) A guide for land drainage engineers on conservation, amenity and recreation. In: *Conservation and Land Drainage*, Water Space Amenity Commission, pp. 36–54.

[38] Drummond, I. (1977) Conservation and land drainage. *Water Space*, **11**, 23–30.

[39] Water Space Amenity Commission (1980) *Conservation and Land Drainage Guidelines and Working Party Report.* WSAC, London.

[40] Newbold, C., Pursglove, J. & Holmes, N. T. H. (1983) *Nature conservation and*

river engineering. Nature Conservancy Council, Peterborough.

[41] Lewis, G. & Williams, G. (1984) *Rivers and wildlife handbook: a guide to practises with further conservation of wildlife in rivers*. Royal Society for the Protection of Birds and Royal Society for Nature Conservation.

[42] Ministry of Agriculture, Fisheries and Food (1988) *Conservation guidelines for drainage authorities*. MAFF/Dept. of the Environment/Welsh Office.

[43] Ministry of Agriculture, Fisheries and Food (1986) *Field drainage and conservation*. Booklet 2522.

[44] Gardner, J. L., Dearsley, A. F. & Woolnough, J. R. (1987) The appraisal of environmentally sensitive options for flood alleviation using mathematical modelling. *J. Inst. Water and Environmental Management*, 1, 2, 171–184.

[45] Turner, K. & Brooks, J. (1988) Flood protection for an environmentally sensitive area: an economic, ecological analysis. *Special Technical Session on Economics of Flood Control and Non-Structural Measures*. Int. Comm. Irrig. and Drainage, Dubrovnik, Yugoslavia, Sept.

[46] Ministry of Agriculture, Fisheries and Food (1985) *Investment Appraisal of Arterial Drainage Schemes*. MAFF, Land and Water Services, River Coastal Engineering Group, London.

[47] Construction Industry Reseach and Information Association (1984) *The retention of wetlands: lowland areas generally defined by gravity*. CIRIA project RP318, Silsoe College, Bedford.

[48] Spoor, G., Chapman, J. M., Hann, M. J. & Leeds-Harrison, P. B. (1988) *Water regime requirements of wildlife habitats*. Paper to MAFF Conference of River and Coastal Engineers, Loughborough, 5–7 July.

[49] Sutherland, D. (1988) *Evaluation of land drainage maintenance on Malltraeth Marsh, Ynys Mons, Gwynedd*. Unpublished MSc thesis, Silsoe College, Bedford.

[50] Morris, J. (1986) Evaluating the wetland resource. *Journal of Environmental Management*, 24, 147–156.

13

Sympathetic approach to river engineering

N. T. H. Holmes, BSc, PhD [†]

ABSTRACT

Rivers are of great ecological importance in Great Britain. However, the wildlife value varies from region to region, partly owing to edaphic conditions and partly to ways in which they are modified and managed by Man. This paper describes some of the features of rivers which make them of wildlife interest as well as highlighting aspects of their interest which have declined owing to unsympathetic management practices. Sympathetic management practices have been attempted in many regions and have been shown to integrate the needs of river engineers with those of both fishery and wildlife conservationists. Examples are given in the paper.

KEY WORDS   River engineering   Wildlife value   Management   Design   Maintenance
            Extinctions   Habitat restoration

INTRODUCTION

Rivers, and the corridors of land through which they flow, are one of Britain's major wildlife resources. The extent of rivers has diminished little throughout the last few centuries, yet their wildlife potential has often been considerably degraded. Rivers differ, therefore, from habitats such as ancient woodlands in being modified rather than totally destroyed. They also span the whole altitudinal and geological range of the British Isles. In addition they serve as corridors which link habitats which have become isolated by intensification of land use.

The importance of rivers as a wildlife resource is illustrated by the diversity of their own environments and those intricately linked with them. Apart from winterbournes, drainage ditches, and feeder streams, the channel will always be a permanently wet habitat of infinite variety. The banks represent habitats of change due to erosion, deposition, water level, fluctuation, etc. Impinging on the river may be an array of different natural, semi-natural, and man-made habitats.

The value of particular river systems for wildlife frequently depends on the degree and type of management within the channel, the extent of development within the corridor,

[†] Nature Conservancy Council.

and the water quality. The natural characteristics of rivers are, in the first instance, governed by their hydrology which is affected by altitude, slope, geology, climate, and many other factors. The range of natural features often represents the maximum potential for wildlife, and it is these which can be degraded or altered by management factors.

Rivers which have been subject to little management usually have a diverse physical structure with variations in substrate types, depth, velocity, and bank morphology. Examples of the range of physical habitats include: flanking rocks, stream bed boulders, waterfalls, rock-pools, shingle and gravel beds, riffles, pools, slacks, shallow earth banks, steep cliffs, muddy edges, islands, fringing swamps and trees. Little-managed adjacent habitats may range from permanently wet bogs, oxbows, fens, carrs, and marshes to drier heaths, scrub, permanent pastures, and woodlands. A bordering strip of trees and shrubs has many important ecological functions and often represents the crowning glory of a natural river landscape.

In contrast, intensively managed river corridors usually have a different character. River and bank morphology may become uniform, replacing a mixture of natural features with simple, disturbed, or artificial structures. Adjacent wetlands habitats may be drained, with intensive agriculture practised right up to the river's edge. This results in little of nature conservation interest and minimal aesthetic appeal. Management practices affect plants and animals in different ways. For example, many wetlands species have become very rare in large parts of Britain after the period of Victorian drainage which diminished considerably our wetland resources. This has accelerated since World War II, from which time the growth of agriculture was prompted through intensive drainage and use of artificial fertilizers. Conversely, some species have thrived in these post-war years. From a nature conservation point of view it is imperative to understand that it is the species which are specialized and dependent on particular environmental features which have declined, while the opportunistic, often 'weed' species, of no specific habitat affinity, have thrived. Numerous wetland plants and animals have declined in abundance, some have become threatened with extinction, and some have actually become extinct. The causes of these declines are often complex, but in many cases the underlying reason is habitat destruction.

Habitat loss in rivers and flood plains occurs in four major ways. Firstly there is the gross degradation of the diversity of channel morphology which may result from over-zealous improvement schemes. Secondly there is the cumulative loss of habitat features on river banks where protuberances and recesses are shaped into uniform batters to reduce rugosity and assist the efficient flow of water downstream. This often results from river maintenance work, but may occur as an incidental downstream product of changes in river management upstream which have altered the flow and sediment patterns of the whole river. Thirdly there is the destruction of natural or semi-natural adjacent habitats for urban or agricultural use; this may occur without necessitating changes to the channel morphology. Fourthly there is the intensification of agricultural use on adjacent land which results from dredging the river; this usually results in a significant lowering of the watertable and may also lead to eutrophication of water bodies.

Nationally, the effects of drainage have been instrumental in the extinction of seven plants and have threatened 27 further species with extinction. The situation is alarmingly illustrated by referring to data for a single lowland county in England, Bedfordshire. In the 180 year period up to 1975 almost 10% of the plant species previously known in the county became extinct. For some habitats such as woodlands, calcareous pasture and waste-land the loss of species has been small. However, marshland species have been lost at an alarming rate: 30% of the flora recorded from marshes in the 18th century have now disappeared from the county. Of the 38 species lost from wetlands in Bedfordshire 40% were lost during the extensive drainage of marshes and fens at the end of the 18th century, 30% were lost in the period 1805–1945 (1 species every 13 years), and 30% were lost in the 30 years between 1945 and 1975 (1 species every 3 years).

It was because of mounting evidence that river improvement schemes and certain maintenance activities had had a generally very bad effect on river and valley wildlife that duties of drainage authorities to consider the environment were strengthened in the Wildlife and Countryside Act. Now there was imposed a 'duty to further... flora and fauna...' so long as it was consistent with their other duties. This amendment went much further than the previous Section 48 of the Water Act 1973 which only required water authorities 'to have regard for' flora and fauna. The present requirement 'to further' conservation applies to water authorities and internal drainage boards in England and Wales, but does not apply in Scotland.

In the past five years it has been clearly shown by many water authorities that it is possible to create extra channel capacity required in capital schemes without detriment to the plants and animals of the river; indeed several schemes have been implemented which have actually improved the environment for wildlife. Improved design for capital schemes has been also matched by modified working practices and techniques for maintenance. Here there have been many examples of management creating improved river habitats in executing works which would have had disastrous consequences for the environment if they had been undertaken in a manner typical of a decade ago. We are now in the era of river management where land drainage or urban flood alleviation objectives can, should, and usually are, integrated with environmental consideration. These latter are, at the very least, that the habitats for plants and animals are maintained after the conclusion of the work, and at best actually better than before the work was initiated.

Such a radical change had been achieved by a much better understanding of river engineering by nature conservationists, and a greater appreciation of the habitat needs of plants and animals by those who manage our rivers. Together they have, through the Nature Conservancy Council, developed a method of surveying rivers so that information on the most desirable approach to river management can be presented to the river engineers. In tandem a close consultation procedure has evolved so that a 'partnership approach' has developed. It would be wrong to herald the end of environmental destruction at the hands of river engineering works; however, the 1980s have witnessed a major shift in emphasis so that those who execute river management work to the detriment of the countryside are in the minority and are often shunned by their enlightened colleagues.

EXAMPLES OF SYMPATHETIC MANAGEMENT

This part of the paper gives a few examples of the manner in which 'river works' can be executed so as to benefit, or at least not harm, the environment of the river and its surroundings. The examples represent only a small fraction of the sympathetic ways in which those engaged with river works are developing ways of executing their duties in the most desirable way for wildlife. The examples include weed cutting, bank maintenance, and channel dredging only; there is not the scope here to include subjects as bank protection, capital designs, etc.

1 Weed Control

Before many modern improvements in channel capacity produced by widening and deepening, the management most practised to reduce the risk of flooding was weed cutting. Traditionally, it is a management technique most associated with chalk rivers and streams, which create the ideal environment for plant growth. Today, however, cutting weeds occurs in virtually all lowland rivers, as well as in artificial drainage channels. The reduction in shade resulting from the removal of flanking trees favours excess plant growth; run-off from the land is now richer in silt and nutrients, which further exacerbates the situation.

The practice on the Test and Itchen is a particularly sensitive example of weed cutting on chalk rivers. Weeds are cut three times a year (April, June, and August), but stretches are cut in strict rotation and with no cutting allowed outside the prescribed times. The Fishing Association informs all owners and fishery interests of the dates as well as issuing reminders of what is expected of them. During the assigned week of cutting in any one stretch the weed is cut by the owners and fishery interests and allowed to drift down river to areas where it is collected by authority staff. Cutting is banned on the last day of the cut in each week and outside the three weekly periods assigned to that stretch. After cutting one stretch of river, another is cut the following week.

This method has numerous advantages for wildlife, as follows:

(1) The period of weed cut is limited to a discrete time of year throughout the river. Staggering these periods in different parts of the river ensures that uncut vegetation is always available.
(2) Staggering sections weekly allows migration of invertebrates and fish to areas of retained cover. The areas cut first in the rotation regenerate to provide cover before the last section is cut.
(3) Removal of weed by the water authority prevents eutrophication and oxygen depletion: cut material degrades on the bank and not in the river.
(4) Food and cover for animals are always available since the maximum area cut at any one time does not exceed the length of an individual beat (usually 2 km).

Cutting weedy lowland rivers and drains

In small watercourses, with a bed width less than three metres, it is often very difficult to retain significant amounts of uncut vegetation if containment of water within the channel is critical. However, in wider ditches, brooks, and rivers some retention of both submerged and emergent plants should be achievable during routine weed cutting operations.

Many authorities no longer stipulate that bank-to-bank cutting is required, instead issuing guidelines for variable amounts of the channel to be left uncut; this is usually dependent on the size of the river and degree of flood risk. Severn–Trent's previous standard approach to weed cutting was to take all weed out of channels in intensively farmed areas. Their recent investigations have shown that this is not necessary, and trials have been set up to see how effective 80 and 60 per cent cuts are in comparison. The authority has also begun to experiment by cutting a meandering path through weeds in straight channels.

The advantages for wildlife which occur as a result of leaving some vegetation uncut are numerous, including:

(a)   Cover is always available to fish and invertebrates.

(b)   Food is maintained for all wildlife.

(c)   Spawning habitats for fish are maintained.

(d)   Nesting sites on reeds are maintained for birds.

(e)   A link is retained between aquatic and terrestrial habitats for insects which have aquatic larvae and flying adults.

(f)   Aquatic insects can use the plants for egg-laying.

(g)   Protected slack water and faster open water are maintained during periods of increased flow.

2 Bank and floodbank maintenance

Banks of rivers provide many contrasting habitats for plants and animals, many of which are dependent on width, aspect, slope, degree of wetness, and soil type. However, whatever the physical nature of the bank, its ecological interest will be changed by different management practices.

A closely and regularly mown grass bank will have little of ecological interest, but might be highly desirable in an urban scene. Regular mowing will reduce the flower content of the bank to a few specialist species regularly encountered on lawns—daisy, buttercup, clover, etc. Since they are rarely left long enough to flower and seed before the next mowing there is little or no chance of insects and other invertebrates making use of these areas for feeding or breeding. The short sward and lack of cover also make such managed banks, even in the absence of human disturbance, totally unsuitable for any mammals or nesting birds.

Banks which are mown only once a year are totally different. Provided that the single mowing is left until late autumn, herbs and grasses have an opportunity to grow to their full height, flower, and set seed. This has many advantages for a very wide range of wildlife interests.

The key to the advantages which result is that the range of flowers and grasses is far greater than in regularly mown banks, or for that matter most banks which are never mown or grazed. This richness of species in turn results in a greater diversity of physical microhabitats, providing cover for many small animals and a greater range of nectar, vegetative shoots, seeds, or fruit for animals to feed on.

Tall vegetation on river banks affords protection to many small mammals, but all will disappear if the vegetation is regularly mown or grazed. The presence of such small animals along the river banks means that predatory birds and mammals will also be attracted. Without the tall vegetation the small mammals will be absent, and without the small mammals the attractive sight of predators will be a very rare occurrence indeed.

Owing to a combination of financial cut-backs and desire to have greater regard for the environment, several authorities have reduced the frequency of bank mowing. In many areas a previous regime of four cuts a year has been reduced to a single cut. This has been regarded as quite sufficient to adequately inspect for vermin damage, and no additional damage has been reported as a result of this practice. Severn–Trent have a research study underway to assess how the vegetation component of such banks changes as a result of the altered mowing regime.

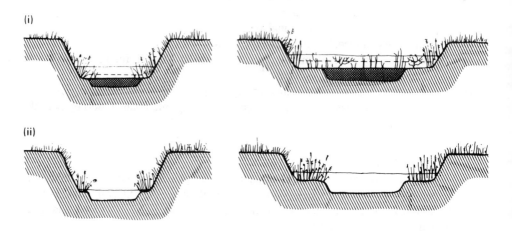

Fig. 13.1.

Where there is sufficient capacity to allow for a small part of the bank to be left uncut, several approaches have been tried. The simplest leaves a strip of uncut vegetation at either the top or the bottom of the bank. There are operational and wildlife advantages and disadvantages with both systems. However, from an ecological point of view, both are valuable for creating bankside habitats for a host of invertebrates, birds, and mammals which would not otherwise use the river if it were mown from top to bottom. If this system is adopted, in combination with an annual mowing of the remainder of the bank, maximum wildlife benefit is ensured.

An alternative method of leaving parts of a bank unmown is selective retention on an area basis. Here banks are left uncut from top to bottom in narrow strips, the intervening sections being mown as normal. Discharge efficiency is only marginally affected since the uncut areas are effectively flattened by floods.

An approach adopted by Wessex is to cut only one bank in any single year. The following year the opposite bank is cut, and the one cut the previous year is left untouched. This has the advantage for land drainage of retaining a good root structure on the banks, adequate facilities for observing structural integrity of the banks, and precludes scrub development. For wildlife it has great ecological benefits since it retains plant richness, stops invasive plants dominating the community, and provides nesting cover for birds and predatory cover for animals as well as a wide range of food in the form of nectar, succulent shoots, fruit, and seeds.

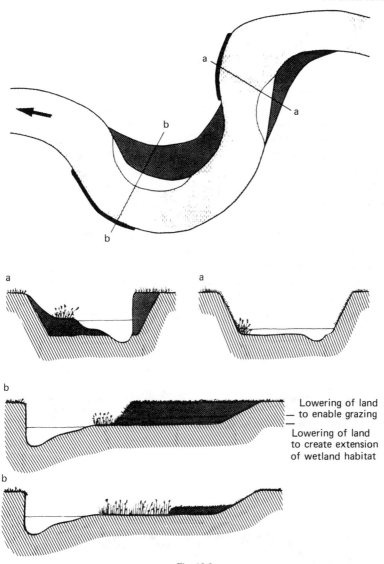

Lowering of land
— to enable grazing

Lowering of land
to create extension
of wetland habitat

Fig. 13.2.

There are two very important operational advantages of this method. The first is that no operator skill is required in identifying patches of particular interest which are most desirable to safeguard. The second is that timing of mowing is less critical than for other systems, provided that breeding birds are not disturbed.

Outside the breeding season, any time is acceptable, since herb-richness is maintained by the seeding which took place the previous year.

3   Channel maintenance dredging
This management activity on rivers has received more attention than any other because of

the detrimental ecological impact it has had on many river systems in past decades. Although channel dredging has the greatest potential to damage existing wildlife interests of a river, it is the river maintenance activity which also offers most enhancement potential.

Figures 13.1–13.4 illustrate just a fraction of the sympathetic methods of dredging at present adopted in routine maintenance dredging. Figure 13.1 shows how retention of some interest can lead to possible enhancement in the future. It shows that any dredging should always leave marginal fringes untouched. In narrow watercourses these will be minimal, whilst in wide rivers they could be 2–5 m wide. These retained edges maintain the present plant and animal assemblage, create a stable 'toe' to the bank, and so reduce future bank-slips or basal erosion, and the undredged edge of the river will be at a different depth from the remainder of the river and, will create a habitat of variation for the future.

Figure 13.2 shows an example of reducing bank erosion, maintaining one of two cliffs, and the creation of a large 'wetland habitat' on the inside of the meander. Despite some initial extra land-take compared with the traditional approach, it is likely that long-term loss of land will be less because a stable solution has been implemented. The advantages for wildlife are manifold compared to the trapezoidal 'eased' bends of old—kingfisher will still be able to nest on the vertical cliff, and brooklime, meadowsweet, sedge, and reed will thrive on the opposite low bank.

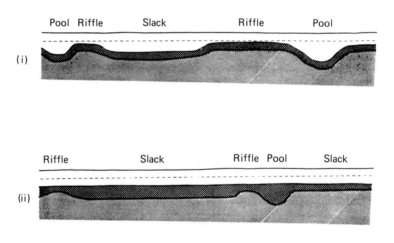

Fig. 13.3.

The importance of variable channel habitat, the pool/slack/riffle sequences, for river plants and animals cannot be overemphasized. Figure 13.3 illustrates the sensitive approach to dredging where such riffle and pool sequences exist. For land drainage purposes it may be sufficient to remove only the 'high spots', but for wildlife this creates a featureless and undesirable mono-habitat. To maintain the slacks and pools it will be necessary to overdeepen these areas. Figure 13.3 also illustrates a method of creating variability where uniformity exists. This variability can be achieved in several ways, the one illustrated being a deliberate attempt to dredge some areas deeper or shallower than the norm. Unless the bed is hard, this is unlikely to work. Alternatives which are more effective include importing stones and gravel to raise the bed level over short stretches or building of groynes to narrow the channel and locally create increased velocity.

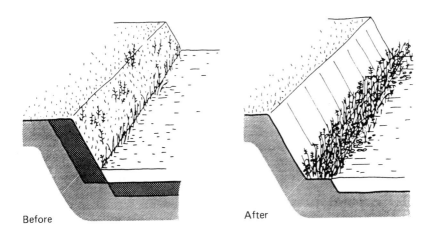

Fig. 13.4.

Finally, Fig. 13.4 shows an example of habitat creation on a bank of minimal wildlife interest. It shows a bank which has been reprofiled with a wet ledge at its base. Examples of this type of enhancement are now very common and have resulted in diverse marginal floras and faunas returning to our rivers which require to be managed for other purposes.

ACKNOWLEDGEMENTS

Parts of this paper are reproduced by kind permission of the Nature Conservancy Council (From: *Surveys of wildlife in river corridors—draft methodology*) and the Institution of Water and Environmental Management (From: *IWEM Manual* No. 8, 1989).

14

Set-aside and extensification of agricultural production: implications and opportunities for nature conservation and the river engineer

C. Newbold, BSc, PhD [†], and A. Rush BA [†]

ABSTRACT

Agricultural overproduction will require some land to be transferred into other uses over the next 10–15 years. Estimates vary, but are likely to be in excess of 1 million hectares. Regulation 1760/87 recently finalized by the EEC provides the basis of several mechanisms whereby farmers will be compensated for taking land out of production. These mechanisms are set-aside in cereals, extensification in cereals and beef, farm diversification, support for afforestation and woodlands, Environmentally Sensitive Areas, and Hill Livestock Compensatory Allowances. This paper discusses the potential of these mechanisms to effect a structured change in land use which could help protect water quality from fertilizer run-off and pesticide use, and perceives changes in drainage practice. Other conservation initiatives relating to these mechanisms are also discussed.

KEY WORDS   Agriculture   Nature conservation   River corridors   Set-aside   Land use change
Pollution   Run-off   Buffer zones

INTRODUCTION

It is now generally accepted that overproduction in Europe of certain agricultural commodities, mainly cereals, will require some land to be transferred into other uses over the next 10–15 years. Estimates of the hectarage likely to be affected vary considerably, ranging from 1 million hectares to as much as one third of all agricultural land. A conservative estimate is probably 3–4 million hectares.

Regulation No 1760/87 [1] recently finalized by the EEC contains a range of measures—covering Hill Livestock Compensatory Allowances, Environmentally Sensitive Areas, and support for afforestation and training; but of most interest for the purposes of this paper is the provision which concerns conversion and extensification of

[†] Nature Conservancy Council, Northminster House, Peterborough, PE1 1UA.

agricultural production. The primary purpose of this provision is to restrict the over-production of agricultural commodities. To this end, in July 1988 Parliament approved the statutory instrument which enabled the Government to compensate farmers for taking arable land out of production. Other provisions this year should include the extensification of beef and cereals. Nevertheless, the Nature Conservancy Council believe that all these agronomic measures should also take account of nature conservation aims, and the opportunities now available should allow for a better integration with agricultural policy objectives.

The effects of modern agriculture on the environment are well known. In 1984 the Nature conservancy Council published *Nature conservation in Great Britain* which set out the losses of wildlife habitat since 1949 [2]. The figures are now widely quoted:

○   95% of lowland unimproved neutral grasslands including herb rich hay meadows;
○   80% of lowland sheep walks on chalk and limestone;
○   40% of lowland acidic heaths;
○   30–50% of ancient lowland woods;
○   140 000 miles of hedgerow;
○   50% of lowland valley fens and basin mires;
○   60% of lowland raised mires; and,
○   30% of upland unimproved grasslands, heaths and blanket bogs.

Not all of this is the effect of agriculture; commercial forestry is another important factor, and building development has also played some part. But the major cause of loss this century has been intensive agriculture, and a successful agriculture is dependent on land drainage.

The Nature Conservancy Council considers that any significant change in agricultural policy which is aimed at reducing the area or intensity of management should take account of the scale of these losses. Protecting the remaining areas of wildlife habitat and re-establishing new habitats should be key elements in any policy change. This would accord with the duty of Agricultural Ministers under section 17 of the Agriculture Act 1986 (3) to balance the needs of agriculture with the conservation of wildlife and the countryside, and be in keeping with the Government's statement in the Department of Environment (DoE) Circular 16/87 (Welsh Office 25/87) [4] of the continuing need to protect the countryside for its own sake rather than primarily for the productive value of the land.

Regulation 1760/87 *Diversion, extensification and set-aside*
There are three principal ways by which production can be reduced, by diverting land into rotational or permanent fallow that is set aside; by reducing intensity so that production per unit area goes down, that is extensification; by diverting land into other uses, that is farm diversification. Examples of the last-named type of diversion are farm woodland schemes and the use of the land for caravan sites and golf courses.

'Extensification' is defined in the regulation as a reduction in the output of the product concerned by at least 20% without other production capacity being increased. It is therefore necessary that some sort of production audit is carried out on farms receiving the aid. A beef extensification scheme should have been introduced on 1 January 1989, but the problems of auditing such a scheme have caused delay in the establishment of the conditions governing the scheme. A cereals extensification should be in force be 31 December 1989, but this may also be delayed.

A set-aside scheme for cereals was laid before Parliament in July 1988, and is now in operation. Farmers are being asked to set aside at least 20% of their arable land, proved to be under production for the season 1987/88, for a period of five years, in return for an annual payment. The scheme is entirely voluntary. Many farmers have registered their arable land—for example 63% of the total arable land of 400 000 hectares has been registered in Scotland. This registration gave the farmer the option to take up the scheme at some future date. In England, Wales, and Scotland some 60 000 hectares of arable land are now committed to the set-aside scheme, of which provisionally 15 300 hectares are in Scotland.

The 60 000 hectares represents 0.01% of the arable land in the British Isles which stands at approximately 6.75 million hectares.

This paper concentrates on the immediate issues of set-aside in arable land and beef and cereal extensification. Clearly, however, agricultural policy changes can be expected to extend beyond these sectors in the near future. This paper should be seen as a broader review of the opportunities for developing a closer integration of agriculture with conservation, and places special emphasis on wetland conservation in or by rivers and drainage channels.

The less favoured areas

The Less Favoured Areas broadly cover the uplands of Britain. Many reservoir catchments are found within the uplands, so land use change brought about by conservation grants could affect reservoir water quality. The regulation as a whole, however, offers little of a positive nature to improve the situation for the L.F.A. farmer (Table 14.1). Nature conservation problems such as overgrazing and inappropriate burning regimes could, in the Nature Conservancy Council's view, be influenced by some restructuring of the hill farming support mechanism and a greater use of beneficial measures like aspects of the Environmentally Sensitive Areas mechanism. The Nature Conservancy Council is concerned to see that the existing structure of hill farming is maintained, but at a less intensive level, a sheep extensification scheme is required. The sale of hill land for afforestation is currently the major factor in the loss of semi-natural habitats in the uplands. Similarly, the water industry is concerned that catchments supplying reservoirs should be protected from afforestation, They have the provision to designate water protection zones for such areas under the Control of Pollution Act 1947 section 31 (5) [5].

Conservation principles

Given the losses which the wildlife resource has sustained from intensive agriculture, set aside and extensification should aim at:

(a) protecting existing areas of semi-natural vegetation from any further loss or deterioration in quality;

(b) similarly protecting and where possible enhancing existing good wildlife habitat;

(c)   encouraging the restoration of wildlife habitats in areas where they have disappeared particularly in areas subject to farming.

Table 14.1. Grants available for set-aside in arable

SET-ASIDE			
Fallow for the 5 year period	Rotational fallow	Non-agricultural, i.e. farm diversification	
		Land	Buildings
1. £200/ha (£80.93/a) payment annually in arrears	1. £180/ha (£72.84/a) payment annually in arrears	1. £150/ha (£60.70/a) payment annually in arrears	1. £150/ha (£60.70/a) payment annually in arrears
2. Green cover crop to be established and cut annually	2. Same conditions as for fallow	2. Permitted uses: Tourism Parking Game-nature reserves Sporting facilities Riding schools	2. Permitted uses: Farm-based industries Farm shops Accommodation Rural Education Livery & Horse stabling (for hire) Rural craft centres
3. Minimum 15 m wide strip of land but must have at least 1 ha (2.47 a) set-aside as whole field	3. Notification to be made annually on area of rotation		3. Planning permission required in normal fashion
4. Notification to be given to landlord	4. Area of set-aside land can only be increased from year to year		

MAIN POINTS OF SET-ASIDE

1. Minimum of 20% of RELEVANT ARABLE CROPS to be set-aside based on 1987/88 cropping, which should have been registered on 21 October 1988. Otherwise independent evidence of this period will be required.
2. 5 year scheme with provision to opt out after 3 years.
3. No agricultural production on set-aside land.
4. Occupier must have farmed land since 1 October in the year before application.
5. All land in less favoured areas (LFA) to receive £20/ha (£8.09/a) lower payments.

Naturalness and semi-naturalness

In Britain there are few truly natural habitats, in the sense of those unmodified by man, and the main reservoir of wildlife lies in the 'semi-natural' areas. This term is applied to plant communities which owe their character to some degree of human intervention, but remain composed of native species and have structural features corresponding to those of natural types. Agriculture, as the predominant land use in Great Britain, exerts a strong influence over the development and survival of such communities. In the lowlands, the main semi-natural areas are now relatively small and fragmented; they include ancient woodland, ancient hedgerow, unimproved grassland, undisturbed fen, unreclaimed saltmarsh and sand dunes. To the west and north of the country, where agricultural use is generally less intensive, considerable expanses of semi-natural vegetation still remain on the moors, mountains, and peatlands. Some of these areas are in part protected as National Nature Reserves (NNR) or Sites of Special Scientific Interest (SSSI), but the remainder must be safeguarded through general countryside policies. It is these latter areas with which this paper is concerned.

Good wildlife habitat and artificial habitat

Plant communities which include some non-indigenous species but still retain a large semi-natural component can be termed good wildlife habitat. Examples are recent broadleaved plantations, recent hedgerow, and partly-improved grassland which retains elements of the original plant communities. Artificial habitats are those composed mainly of non-indigenous species—arable fields, grass leys, conifer plantations.

All these definitions are to some extent arbitrary classifications. In practice the range from 'natural' to 'artificial' is a continuum; and nature conservation is especially concerned with the upper part of the range.

Protecting semi-natural and good wildlife habitat

Generally speaking, these habitats form the least productive part of any farm so they are unlikely to be the direct focus of set-aside of extensification measures. However, they may be at risk from any restructuring of the farm business consequent upon entering a scheme, for example reducing the cereal area on a mixed farm may encourage the farmer to build up the livestock elements of the business and bring semi-natural grassland into more intensive use. This could be guarded against by a simple farm development plan which identifies the areas of semi-natural and good wildlife habitat on which intensive management techniques should be avoided.

An important principle of extensification should be a re-examination of all other agricultural schemes to ensure that they are not damaging to nature conservation interests. There is no justification for any further destruction of semi-natural and good wildlife habitat for the purposes of agricultural production. The Agriculture Improvement Scheme should therefore be reviewed in order to phase out grant-aid for activities which can be damaging to wildlife habitats, land drainage grants being a particular case in point. These should be replaced by new schemes which provide farmers with incentives to retain wildlife habitats and manage them sympathetically. This could entail an extension of some of the mechanisms now available to farmers in Environmentally Sensitive Areas (ESAs).

Re-establishing wildlife habitat

The potential for this in the farmed countryside is very large, and opportunities should now be taken to restore in some measure what has been lost over the years. But it must be emphasized that creating new habitat is not a substitute for conserving what still remains; the new habitats may never be as rich or ecologically varied as existing semi-natural areas. In addition, it is clearly more cost-effective to retain existing habitats than to create new ones. That said, the new habitats are valuable in their own right, and if created on a more considerable scale than hitherto, could contribute greatly to the diversity and visual amenity of the countryside.

Any set-aside or extensification scheme must therefore contain two elements: the protection of existing areas of semi-natural and good wildlife habitat, and the creation of new habitats.

The set-aside scheme in arable areas and its ecological implications

The present set-aside scheme does little to encourage farmers to protect remaining semi-natural and good wildlife areas. There are no provisions for a farm plan which would identify such wildlife areas on which intensive management techniques should be avoided.

There is yet another major disadvantage of the UK set-aside scheme. The UK Government has decided that the set-aside land cannot be grazed, although the EC Directive included the option of set-aside land being turned into grassland for grazing animals. Our Government has succumbed to pressures from livestock producing areas and were particularly concerned that incomes from sheep farming should be maintained in hill areas.

There are also no provisions within the annual payment whereby habitats can be re-created or enhanced.

There are three elements to the set-aside scheme:

(i)   payments are made for fallowing; lower payments are made to L.F.A. farmers for both rotational and permanent fallow. In all cases, permanent fallow, that is for the five year period, attracts a higher grant (Table 14.1);

(ii)   the farm can be diversified for permitted uses such as tourism, parking, and nature reserves, but these attract a much lower grant (Table 14.1);

(iii)   land can be planted up as woodland but it must be tied into the Farm Woodland Scheme or the Woodland Grant Scheme (see Table 14.2)

Fallows

Whole farms, whole fields and field margins 15 metres or more wide can be taken out of production, but all should be maintained in good agronomic condition. If there are serious infestations with weeds such as sterile brome, couch, and black grass these should be eradicated through ploughing or, with a special dispensation, the use of herbicides. Cover crops are likely to be used to suppress these weeds and low productivity grasses of potential conservation value and conservation seed mixtures could be used but the annual payment will not cover the additional cost. The value of such mixtures is in doubt anyway since to maintain the sward, such areas need to be grazed or cut annually and the cuttings removed. The present scheme only allows cutting but does not allow removal. Any enhancement work such as the planting of new hedgerows or the filling in of hedgerow gaps, whilst allowed, is not compensated for under the annual payments.

The DoE are seriously considering a 'top up' scheme whereby farmers are paid an additional sum so that seed mixtures can be purchased, hedgerows can be re-planted and general enhancement work carried out.

Farm diversification, or non-agricultural use under set-aside

Farm diversification or non-agricultural use will allow agricultural land to be used for schemes where the area is not kept in good agronomic condition (Table 14.1). Land could be used to create nature reserves, but the payments at £150/ha are insufficient to attract many farmers. Much relies on the 'top-up' scheme.

Farm woodlands

The Farm Woodland Scheme and the Woodland Grant Scheme allied with a set-aside payment provides a farmer with enough incentive to convert his arable land to conifer or broadleaved woodland. The package implies a 10–40 year contractual arrangement with the Forestry Commission.

Table 14.2. Grants available for set-aside in arable

WOODLANDS with-set-aside	
WOODLAND GRANT SCHEME (WGS)	WOODLAND GRANT SCHEME plus FARM WOODLAND SCHEME (FWS)
1. £200/ha (£80.93/a) annually in areas depending on land classification plus WGS payments	1. FWS payments: £190/ha (£76.89/a)—Normal £150/ha (£60.70/a)—Disadvantaged areas £100/ha (£40.47/a)—Severely disadvantaged areas

Area approved (ha):

Area approved (ha)	Conifers	Broadleaved Trees
0.25–0.90	£1005	£1575
1.0–2.90	£880	£1375
3.0–9.90	£795	£1175
10 +	£615	£975

2. Duration:
   40 years—oak & beech
   30 years—broadleaved & mixed
      woodland containing more than
      50% broadleaved trees
   20 years—other woodland
   10 years—traditional coppice

2. Payments
   70% on completion of planting
   20% after 5 years
   10% after 10 years

3. Payments through WGS grant:

Area approved (ha)	Conifers	Broadleaved trees
1.0–2.90	£505	£1375
3.0–9.90	£420	£1175
10+	£240	£975

3. Approval through Forestry Commission
4. Landlord's consent required

4. Payments in similar 3 installments
5. Approval through Forestry Commission
6. Landlord's consent requirement

Conservation set-aside

Nature conservation clearly falls within the 'non-agricultural uses' to which land may be diverted. Substantial opportunities exist for re-creating habitats on land taken out of cereals, whether the areas withdrawn are whole farms, single fields, or strips alongside field margins.

Fields and farms

Specialized techniques could be used for re-establishing a range of habitats in areas which they have disappeared. The main opportunities are:

(a) Creation of herb-rich grasslands by using commercially available conservation seed mixes. This technique is applicable in a wide variety of situations and soil types.

Once established, the grasslands can be maintained by using low intensity management techniques.

(b)   Fields in some areas where goose grazing is a problem could be set aside and sown with grass to provide winter feeding grounds. This would reduce damage to adjacent arable fields which would continue to be intensively managed problem is highlighted, within the Practical Guide on the scheme.

(c)   New woodland could be planted with native species for a mix of conservation, sporting, and amenity objectives with some timber production, for example coppicing for fuel wood. Scrub and woodland could also be allowed to regenerate naturally to establish, at low cost, a resource of timber for pulp and other purposes.

(d)   Certain carefully selected areas could be allowed to develop their natural climax vegetation with minimal or no management intervention. In areas of countryside dominated by intensive management the establishment of such 'wilderness areas' would add greatly to interest and diversity, and could form a valuable experimental and educational resource.

(e)   In traditional sheep-rearing areas like the chalk downs, reversion from arable to pasture would be valuable, though care must be taken to avoid competition with the Less Favoured Areas. The aim should be for very low intensity systems with zero fertilizer input, compatible with maintaining a species-rich sward. This option is not strictly 'extensification' within the terms of the Regulation, but could be considered as a form of conversion, the prime purpose being nature conservation rather than sheepmeat production.

(f)   In re-creation of heathland in areas where it has declined, especially Dorset, Hampshire, and Breckland, the best results will be on light acid soils where nutrient levels can more easily be reduced. Proximity to surviving areas of heathland which can act as a seed source is essential, although as an alternative technique conservation seed mixes containing a specifically heathland flora can be sown.

(g)   To meet set-aside and drainage standards, a structured reduction in drainage standards in appropriate areas could be used to re-instate washlands, or to create temporary or permanent wetland habitats for wintering wildfowl and other wildlife. The wildlife value of river corridors could be enhanced by this technique.

Conceptually, there is little reason why farmers could not be persuaded to set aside arable land alongside stream and river banks to produce an extensive and much enlarged river corridor. Even large areas of arable on flood plains could be taken out of production and drainage standards restructured. Gaps in the corridor not under arable could be turned into woodland by using the woodland grant scheme without set-aside (Table 14.3). Such a structured approach to an uninterrupted river corridor could also be an important natural filtering mechanism or buffer zone against agricultural pollution.

(h)   With regard to set-aside and water pollution, the surface waters of whole catchments could be protected from the excesses of silage liquor run-off, animal slurry pollution, and pesticide and fertilizer run-off by using a corridor to absorb these pollutants. This concept would protect only against diffuse sources unless potential point sources of pollution were also buffered.

Table 14.3. Grants available without set-aside

WOODLAND GRANT SCHEME plus FARM WOODLAND SCHEME	DO NOTHING Without set-aside WOODLAND GRANT SCHEME WOODLAND GRANT SCHEME plus BETTER LAND SUPPLEMENT		
1. Payments similar to those under set-aside	1. WGS payments:		
	Area Approved (ha)	Conifer	Broadleaved Trees
	0.25–0.90	£1005	£1575
	1.0–2.90	£880	£1375
	3.0–9.90	£795	£1175
	10 +	£615	£975
2. Minimum size 3 ha (7.41 a) per holding with minimum planting size 1 ha (2.47 a)	2. Payments: 70% on completion 20% after 5 years 10% after 10 years		
3. Maximum planting size 40 ha (98.8 a) per holding	3. Additional ONE OFF PAYMENT: of £200/ha (£80.93/a) on arable and improved pasture		
4. Planting can take place on arable land or improved pasture			

Tables 14.1–14.3 are reproduced by kind permission of Savills plc, 20 Grosvenor Hill, Berkeley Square, London W1X 0HQ.

With the UK Government's reticence to designate water protection zones using corridors of former agricultural land could be one way forward in protecting rivers such as the Taw and Torridge in Devon from agricultural pollution.

Agriculture pollution has become such a problem in the state of New Jersey, USA, that a law has just been enacted which protects those wetlands surrounded by agricultural land with buffer zones. The width of the buffer zone required is calculated by using a mathematical model which computes such factors as soil type and slope.

The Dutch government has placed a similar bill before parliament, and such methods have been used for more than a decade by the New Zealand government [6, 7].

Set-aside and sea-level rise

The gradual rise in sea level also presents many wildlife opportunities, allowing marginal agricultural land to revert to salt marsh and other habitats. The cost of maintaining an expensive sea wall to protect unwanted agricultural land could well be prohibitive. A valuable wildlife habitat could be restored, and the salt marsh would act as a soft form of coastal defence, absorbing wave energy before the waves hit the 'new' sea wall now placed further back. Coastal defence lines need not be constructed from new. Former sea walls, now inland, could be repaired and used as a cheaper option for coastal defence. A rise in sea level could also see the need for wider channels inland or an increased bank height. One option would be to construct new washland areas.

Field margins and set-aside

The primary aim under a fallow system should be the establishment of rough grass margins, 15 metres wide, around the perimeter of intensively managed arable and grass fields. The margin could be established by sowing a suitable seed mix or allowing the rough grassland to develop naturally. No inorganic fertilizers, slurry, or pesticides can be applied to the grass margins, so that plant communities characteristic of rough grassland would develop.

Hedgerows offer another specialized method of field margin management which help to reduce agricultural productivity and restore wildlife habitats. 'Free growth' of hedges could be permitted by encouraging them to grow upwards and outwards without cutting. It would be valuable if some hedgerows could be allowed to develop into bands of scrub, which would ultimately grow on to woodland. Fences should wherever possible be replaced by hedges, with the help of Agriculture Department grants.

Another specialized type of field margin, suitable in some areas of relatively wet ground, would be the establishment of reed (*Phragmites*) fringes. These could be planted between the crop and the ditch to provide new habitat and to intercept surplus fertilizer.

The above two types of specialized management of field margins would have conservation rather than reduced agricultural production as the primary aim, but would, nevertheless, make a contribution to reducing production (and fertilizer run-off) if introduced on a wide enough scale.

Creating a network of rough grassland and other habitats along field margins may be more beneficial for nature conservation than taking out isolated whole fields. The existing margins are often less intensively sprayed with herbicides and pesticides, and in consequence contain remnant populations of native plants. These will more readily colonize the new areas of rough grass and more specialized types of margin than the centres of large fields remote from the nearest seed sources. Moreover, the network of margins would provide a series of habitat 'corridors' or 'stepping stones' between existing areas of wildlife habitat, offering further opportunities for insects, small mammals, and plants to extend their range.

Where measures are being taken to establish field margins or fields of low intensity management on a farm, opportunities should be taken to locate these near or adjacent to existing areas of high nature conservation value. Not only will this aid the colonization of native species into the new habitats but it will also help to buffer the existing wildlife habitats against the effects of spray drift and pollution of water courses.

Fallowing

Fallowing is the major alternative use for diverted land specified in the Regulation. However, leaving fields fallow in general brings negligible wildlife benefits, given the management entails cultivation and sowing a cover crop—grass or a green manure—to prevent excessive nitrogen leaching, and also in some areas to guard against soil erosion. An exception would be if fallows were allowed to develop an annual or biennial flora, with a mowing regime but no herbicide or fertilizer applications, and ploughing restricted to the end of the rotation.

Other measures to reduce cereal production

As well as land diversion, general measures to reduce the intensity of cereal production would be valuable for nature conservation. Lowering the levels of pesticides or fertilizer input, especially applications of nitrogen (to grasslands as well as to cereals), could help to reduce the pollution of wet meadowland, streams, and river systems.

Reductions in the winter sowing of cereals would also be beneficial in areas where soil conditions allow for spring sowing, leading to lower overall levels of fertilizer and pesticide application. Spring sowing produces a less dense crop, more suitable conditions for ground nesting birds, better feeding for overwintering birds, and more opportunities for wild plant survival through improved capacity for seed shedding. Spring sowing of cereals would be particularly advantageous if combined with the specialized management of the field margins developed by the Game Conservancy. Undersowing with clover or other similar crop would help to reduce problems of nitrogen leaching in the autumn.

Beef extensification

It is recognized that particular problems in achieving surplus reductions in this sector are likely to be encountered by all Member States, for a variety of reasons, including the interrelationships with the dairy sector and the difficulty in policing headage limitations. In Great Britain it appears likely that the 20% reduction will be targetted on lowland suckler herds, and that no reductions will be sought in breeding herds within the Less Favoured Areas.

The Nature Conservancy Council is concerned that the scheme should not encourage too great a depletion in beef breeding herds. Production of suckler calves is often compatible with conserving surviving areas of semi-natural grassland, since the intensity of management is usually less than in dairy or beef fattening enterprises. Moreover, it is important that the supply of store cattle to lowland farms be maintained, since the nature conservation value of certain areas, such as the East Anglian grazing marshes, is dependent on the continuance of a low-intensity grazing regime. It would be preferable if some way could be found of limiting the production of beef calves from dairy herds.

Assuming that the problems of setting up and policing a beef extensification scheme can be overcome, the main potential benefit for nature conservation will lie in a lower intensity management regime for grassland habitats through lower stocking rates, if this is combined with reductions in fertilizer and herbicide applications. The aim should be to lower the intensity of grassland management over the whole of any farm entering the scheme, with particular emphasis on the rough grazing element, and on other areas of permanent grass where semi-natural plant communities remain, whether in complete or fragmentary form. Simple guidelines could be produced for the identification and management of grasslands of nature conservation interest, which could be applied by farmers entering the scheme. The previous remarks on hedgerow management also apply.

Some farmers joining the scheme may wish to seek alternative productive uses for their 'spare' grassland. The terms of the Regulation will not permit other agricultural use, but afforestation under the Farm Woodlands Scheme would be a possibility, as would development for other purposes (tourism, recreation, housing). This could be disadvantageous for nature conservation, since such development is likely to be targetted on rough grazing and enclosed permanent pasture, which generally are of value for wildlife. Opportunities to enhance the nature conservation value of the remaining grassland may also be lost, as stocking rates would not fall (and may even rise). But afforestation could still be of value provided that it was located on the grass leys and areas of improved pasture with little existing wildlife interest.

Timescale

Short-term diversions of land can be valuable for nature conservation, providing habitats for common plants and butterflies, and many farmers will not wish to commit their land to a non-agricultural use for long periods. However, some diversions of necessity will be long-term or permanent, for example if the land is afforested or developed for housing. Land which is set aside for nature conservation will generally increase in interest over time; this applies particularly when the aim is to re-create a habitat such as heathlands, but also applies to the more straightforward process of establishing rough grassland along river margins. It would be valuable therefore if additional incentives could be offered for diversions for ten years and over.

CONCLUSION

It is generally accepted that overproduction in Europe of certain agricultural commodities will require some land to be transferred into other uses over the next 10–15 years. Regulation No 1760/87 finalized by the EEC contains a series of measures to reduce agricultural production. The Nature Conservancy Council believe that all these agronomic measures should also take account of nature conservation aims, and the opportunities now available should allow for a better integration with agricultural policy objectives.

A strategy to sustain and enhance the wildlife resource would aim at:

(a) Protecting all existing areas of semi-natural vegetation from any further loss or deterioration in quality.
(b) Similarly protecting and where possible enhance existing good wildlife habitat.
(c) Encouraging the restoration of wildlife habitats in areas where they have disappeared particularly in areas subject to farming.

In particular, river protection zones using corridors of former agricultural land could be one way of protecting watercourses from agricultural run-off. Wetland riparian zones could be restored in a structured way, using the mechanisms enshrined in Regulation 1760/87 which allows land to be taken out of agricultural production.

REFERENCES

[1] European Commission Regulation (1987) Council Regulation 1760/87 as regards agricultural structures the adjustments of agriculture to the new market situation and the preservation of the countryside.
[2] Ratcliffe, D. A. (1984) *Nature conservation in Great Britain*. Nature Conservancy Council, Peterborough, 112.
[3] Agriculture Act (1986) Her Majesty's Stationery Office, London.
[4] Department of the Environment (1987) *Development on agricultural land*. Circular 16/87 (DoE), Circular 25/87 Welsh Office, Her Majesty's Stationery Office.
[5] Control of Pollution Act (1974) Her Majesty's Office, London.
[6] McColl, R. H. S., White, E. & Waugh, J. R. (1975) Chemical run-off on a catchment converted to agricultural use. *New Zealand Journal of Science*, **18**, 67–84.

[7]   McColl, R. H. S. & Gibson, A. R. (1979) Downslope movement of nutrients in all hill pasture, Taita, New Zealand. *New Zealand Journal of Agricultural Research*, **22**, 279–286.

Discussion on Papers 12, 13, and 14

Mr J. D. Briggs (British Waterways Board), opening the discussion, said that he was a biologist working with, and interested in, nature conservation on waterways. He felt that there had been great changes in management attitude since the 1981 Wildlife Act. With the arrival of 'set-aside' there were further new aspects to waterway and general environmental management.

A keyword was 'imagination', and he considered that if this was used more in environmental management there would be much benefit to conservation. Referring to the paper by Mr Holmes, he said that this illustrated the advent of a more imaginative approach. Weed-cutting was no longer total, emergent vegetation was left where possible, and channel dredging was more sensitive; management in general was more environmentally acceptable.

The engineering tradition of environmental neatness had diminished; it was now realized that habitat destruction could be minimized without affecting engineering functions. It was regrettable that such changes could not have occurred 20 years ago instead of just during the past 5 years or so.

Conservationists were also changing; in the past there had been much to learn from engineering management, and there was now more communication. However, was there a need for a better forum to present the advantages of sympathetic environmental management?

The river corridor studies mentioned by Mr Holmes comprised a vegetation mapping system that could be annotated to highlight areas in need of conservation or enhancement. These annotations could be interpreted by the river engineer. He asked Mr Holmes how far this idea had developed and been accepted by the water and drainage authorities, and whether it was successful in promoting good management?

With reference to Mr Morris's paper, Mr Briggs said that field drainage still seemed to be left to the discretion of the farmer. He asked what controls there were on such drainage? He said that the paper referred to a scientific and traditional approach to land drainage, and asked how much of each was carried out?

All three papers had touched on the topic of habitat creation; a concept that might be the conservation tool of the future. It was relatively easy with water habitats, compared with, for example, ancient woodland or upland heath. With water systems there was the possibility of creating almost natural habitats. He felt that was very exciting, and with 'set-aside' and more sympathetic approaches to river management the water industry should look forward to environmental progress.

Mr L. Woodward (Elm Farm Research Centre) said that Dr Newbold had said that, at some point in the future, market forces would take over and replace grant aid from the

Government or the EC. He asked for expansion on this aspect because he could not see the way that market forces could encourage or pay for farmers to meet conservation or environmental objectives.

Mr B. H. Rofe (Rofe, Kennard and Lapworth), with reference to Dr Newbold's paper, said that he felt that he must question the concept that change meant destroy. He said that change could often mean enhancement, and he felt that there was a danger that conservation was taken to mean that everything that was there must be kept. Very often that which was there was not necessarily the best thing in that particular locality, and it was possible to create something better. He thought that change was opportunity not destruction.

In defence of river engineers, Mr Rofe said that they were the people who thought out the principles of river regime, which were established, particularly in India, almost seventy years ago. He said that these principles had to some extent been rediscovered, but in the period immediately post-war there was tremendous pressure to increase agricultural production. Engineers, as always, were servants of those directing, and had produced the answers that were required at that time. He said that many schemes had been produced with tremendous sympathy—in fact he felt that this had been the majority of schemes. He felt that credit should be given.

Mr R. M. Walls (West Hampshire Water Company) said that previous reference had been made to the number of species that had been lost in lowland UK, particularly during recent years, and he wondered how much one should positively replace those by re-introduction, or possibly even introduction, to other places, to retain a reasonable number of plants and animals if the habitat was suitable.

There was the problem of simply reseeding, because one needed to keep some sort of vegetation on the bank, either in the river or out of it, to stabilize it. This had caused debate in other circles.

Mr S. W. Bailey (ADAS/MAFF) said that Dr Newbold had mentioned straw incorporation and said that this was often recommended as a method of reducing soil nitrate levels in the autumn, but only in the short term. It was recognized that during the first few years it would use a small amount of the surplus nitrogen available in the autumn, but by no means all, so that there was still a risk of leaching. However, straw incorporation also built up the soil's organic matter pool, and in the long term this might exacerbate the problem.

He said that there had been good interest in the previous farm waste grant scheme, and that had also applied to the new scheme for farm waste management facilities.

With reference to barrier zones, Mr Bailey said that he doubted if these would make a significant contribution to solving the problem of water pollution from farm wastes. He said that the annual WAA/MAFF report on pollution from farm wastes showed water pollution by land run-off was one of the minor causes of pollution.

Mrs V. J. Hamlin-Wright (Stratton Streles Estates Ltd) asked if the NCC, or any other organization, ran a counselling or advisory service available to farmers, and if so how much did it cost and was it advertised?

Authors' replies to discussion on Papers 12, 13, and 14

Dr Newbold, in response to Mr Woodward, said that the USA was now putting pressure on the EC to abide by the general agreement on tariff and trade (the GATT), for example last year Spain had to accept an import of wheat which it did not really want. The farming industry within the EC would have to reduce or lose some of its subsidies because the GATT was talking about fair trade generally. He said that the other aspect was on the overproduction side, in that new grains were becoming available to farmers, whereby the cereal production could increase by 6% per annum over the next five years. That would have an effect on the amount of land to be taken out of production. He thought that market forces would operate some time in the future whereby the subsidies would begin to end. That land which was taken out would be clearly the most unproductive land at that time, so there was no way that it could be farmed economically. In some cases, the poorer land was already designated for set-aside, particularly in Scotland, that went into arable use 5–10 years ago. In other cases it was the problem of access or soil erosion which reduced their economic value. If it was accepted that land was going to come out of production and there was no grant to farmers, then the simple answer was that even abandoned land could actually restore quite a lot of habitats. However, he thought that much of the land would go into afforestation because of the woodland grant scheme and the fact that the EC as a whole was not in surplus in timber production.

Answering Mr Rofe, he said that what he had been trying to say in terms of the recorded habitat losses was that in those habitats which had been lost, it would take centuries to re-establish the communities that had been destroyed. He said that he referred either to very diverse communities that were very interactive or to very stable communities. Clearly much ancient semi-natural woodland had been lost—again a long period was needed for restoration. Why should time be spent on attempts to recreate what little was left instead of trying to preserve what remained.

Also, in response to Mr Walls, he said that with regard to reintroduction and reseeding in those areas already lost, the NCC welcomed that with certain provisos. He stressed that it was important to reintroduce something which was fairly natural to the particular region. Reseeding also should provide a base community of low productivity plants by which the natural recolonization of plants could occur.

Responding to Mr Bailey he said that he agreed about the organic build-up, but as he understood it the major nitrate loss occurred when the land was reploughed for resowing in the autumn, and with the autumn rains. He said that some soils were now very impoverished of organic matter and it would take a long time to reach 'saturation'; loss of nitrogen from the organic loadings within the soil would he thought be minimal.

With regard to buffer or barrier zones, Dr Newbold said that he was interested in attacking the problem of diffuse sources of pollution, which he felt were the main problem from the point of view of control because the actual point sources, such as silage stores, should be addressed by COPA. The point sources could be controlled by treatment methods. He felt that there was under-estimation of the diffuse source problem.

In response to Mrs Wright, he said that clearly ADAS was one of the major areas of advice to farmers on conservation matters, and he also commended the Farming Wildlife Trust. He said that both were very sympathetic to farming needs and were free.

In conclusion, with reference to the farm grant scheme, he said that many farmers had said that they could not afford the 50% contribution. He hoped that this would not be the case, but felt that some of the poorer farmers, particularly in the hill areas, would not be able to afford it.

Dr Holmes, in reply to Mr Briggs, said that he believed that the general situation was good, but there was no room for complacency. By continuing and developing the consultation processes, and building up even better expertise on both sides, things would improve further. In general it was now regarded as the only right and proper way of conducting river engineering works, and people were working together to make things work from both sides. River corridor surveys were generally executed by all water authorities before any use of a dredger. Further than that, some authorities not only did surveys before heavy maintenance work and capital work, but actually sent surveyors in to find out more about their watercourses so that they were actually using the corridor survey information to decide what was an appropriate strategy for the management of the river.

Dr Holmes concluded by stating his appreciation of how engineers had been amending their working practices to take account of environmental issues. The environmentalists needed to state what was required to sustain the interests which they wanted to retain, or the kind of habitat that wanted creating, and the engineers should be left to use their expertise in deciding how best to achieve those aims.

Mr Morris said that, with regard to market forces, he felt that whether or not they would encourage or discourage farmers to meet environmental standards depended on which area of farming was considered and whether a particular area was one where farming was already a priority, or one where environmental issues were more important. In intensively farmed areas he suspected that a movement toward market prices would encourage farmers to regard drainage investments as an insurance policy, and that they would be interested in protecting their incomes. He said that on average about 70 000 hectares/annum were drained on a continuing basis, of which most was probably uprating and rejuvenating old schemes. Now that grant aid was not available, there were only about 30 000 hectares/annum put in, but this was still substantial. However, it tended not to be in the lowland river areas but in the intensively farmed areas.

With regard to the general agreement on trade and tariffs, he said that he could not foresee a move towards free trade in agricultural produce on an international basis. Free trade meant reducing protectionism with respect to access to markets, and because of the very close relationship between social policy (rural) and agricultural policy, he could not see governments in the developed or developing world leaving agriculture to market forces. There was likely to be intervention on the domestic agricultural policy front, although 'lip service' might be paid to free trade.

He said that the 1947 Agricultural Act had been a major piece of legislation in which there was a decision to produce that part of the national food which was deemed to be in the national interest. The degree of self-sufficiency that had been achieved in indigenous foods was quite substantial. That was the policy framework, and river engineers and agriculturalists had responded to it. He agreed that river engineers were not there to make 'decisions' but to make designs and recommendations, and offer them for decision-makers who would invariably be politicians. It was important to present such information so that the best decisions could be made. That did not remove all criticism from river engineers or agriculturalists because there was much evidence that during the 1960s and 1970s schemes which were designed for agricultural purposes often had arbitrary and fairly optimistic appraisals. He felt that in some cases, the river engineers had not really known what they had been designing the schemes for, and were not close enough to the farming in those regions.

In conclusion, Mr Morris said that field drainage had been going on for many years and there were many regional variations in drainage practice in the UK. These had evolved over time by farmers and more recently by contractors. He said the scientific approach could be usefully applied to determine a required level of service to farmers so that they were not constrained unnecessarily by poor drainage conditions. The scientific approach was also necessary to enable more consistent and objective decisions about management. The current work by his College with the Severn–Trent WA and the Welsh WA was an attempt to draw up a database which described farming conditions within a catchment, to collate information which determined drainage conditions in a given catchment, and on that basis present information to decision-makers. He said that he would like to see environmentalists applying a similar approach to set alongside agricultural databases, and presented in a way to enable correct decisions to be made. He felt optimistic about the opportunities of working together.

Index